AF361927

Natural and Artificial Unsaturated Soil Slopes

Natural and Artificial Unsaturated Soil Slopes

Editors

Roberto Vassallo
Luca Comegna
Roberto Valentino

MDPI • Basel • Beijing • Wuhan • Barcelona • Belgrade • Manchester • Tokyo • Cluj • Tianjin

Editors
Roberto Vassallo
School of Engineering
University of Basilicata
Potenza
Italy

Luca Comegna
Department of Engineering
University of Campania "Luigi
Vanvitelli"
Aversa
Italy

Roberto Valentino
Department of Chemistry, Life
Sciences and Environmental
Sustainability
University of Parma
Parma
Italy

Editorial Office
MDPI
St. Alban-Anlage 66
4052 Basel, Switzerland

This is a reprint of articles from the Special Issue published online in the open access journal *Geosciences* (ISSN 2076-3263) (available at: www.mdpi.com/journal/geosciences/special_issues/unsaturated_soil_slopes).

For citation purposes, cite each article independently as indicated on the article page online and as indicated below:

LastName, A.A.; LastName, B.B.; LastName, C.C. Article Title. *Journal Name* **Year**, *Volume Number*, Page Range.

ISBN 978-3-0365-1876-3 (Hbk)
ISBN 978-3-0365-1875-6 (PDF)

Contents

Preface to "Natural and Artificial Unsaturated Soil Slopes"

Recently, there have been extensive studies on engineering problems involving soils whose mechanical and hydraulic properties are strongly influenced by the degree of saturation. Earthen embankments, soil–vegetation–atmosphere interactions, geoenvironmental applications, and risk mitigation are just a few examples of the constant interest of the scientific community to the subject. The presence of a sloping ground surface is common to many of these problems. In slightly inclined natural slopes, susceptible to deep landslides, the unsaturated condition of shallow soil horizons affects deep pore water pressure distribution and, consequently, the stability of the entire slope. The stability of steep mountain areas covered by shallow deposits is often guaranteed by a shear strength contribution related to the unsaturated nature of the soil. In this case, the degree of saturation plays a key role in determining which rainfall events can act as landslide triggers, therefore controlling the post-failure evolution. Partial saturation is the basic characteristic of soils adopted as construction materials of geo-structures such as levees, dikes, and dams. This condition governs the structure behavior during construction phases, in serviceability, and in extreme scenarios. Hoping to provide a bridge between theoretical research and practical applications, this Special Issue collects quality contributions related to natural and artificial slopes under unsaturated conditions, and focuses on several aspects, including: water retention and transport properties, mechanical behavior, advances in experimental methods, laboratory and in situ characterization, soil improvement, field monitoring, geotechnical and geophysical field tests, landslide investigation and prevention, the design and maintenance of engineered slopes, analysis at different spatial scales, and the constitutive and numerical modeling of chemo-hydro-mechanical behavior.

Roberto Vassallo, Luca Comegna, Roberto Valentino
Editors

Editorial

Editorial of Special Issue "Natural and Artificial Unsaturated Soil Slopes"

Roberto Vassallo [1,*], Luca Comegna [2] and Roberto Valentino [3]

[1] School of Engineering, University of Basilicata, 85100 Potenza, Italy
[2] Department of Engineering, University of Campania "Luigi Vanvitelli", 81031 Aversa, Italy; luca.comegna@unicampania.it
[3] Department of Chemistry, Life Sciences and Environmental Sustainability, University of Parma, 43124 Parma, Italy; roberto.valentino@unipr.it
* Correspondence: roberto.vassallo@unibas.it

Citation: Vassallo, R.; Comegna, L.; Valentino, R. Editorial of Special Issue "Natural and Artificial Unsaturated Soil Slopes". *Geosciences* **2021**, *11*, 334. https://doi.org/10.3390/geosciences11080334

Received: 5 August 2021
Accepted: 6 August 2021
Published: 7 August 2021

Publisher's Note: MDPI stays neutral with regard to jurisdictional claims in published maps and institutional affiliations.

Natural and artificial slopes are frequently constituted, at least in part, by soils in unsaturated conditions. Their degree of saturation strongly influences the hydraulic and mechanical properties and, therefore, pore water pressures and stability. This Special Issue collects six papers focused on unsaturated soil behavior in engineering applications involving natural or artificial slopes. All the papers hold a significant laboratory/site experimental characterization and/or numerical analysis content for the interpretation of the considered phenomena.

Ponzoni et al. [1] report water retention data for a sandy silt and a silty sand coming from two different climatic European environments. The soils were tested in the laboratory by using two commercial devices, and the results were modeled with van Genuchten's law. The authors determined a correlation of the fitted parameters with the amount of fines and discussed the advantages and limitations of the proposed approach on the basis of drying–wetting test results. The study demonstrated the feasibility of the approach, aiming for a rather inexpensive characterization of retention properties for the assessment of climate impact on slope stability, and providing a careful evaluation and an elaboration of data.

Bovolenta et al. [2] discuss the reliability of soil moisture capacitive sensors, the operations related to their calibration and the installation phases for several European monitoring networks designed to assess the susceptibility to rainfall-induced debris and earth slides. The aim of such networks is to establish a cause–effect relationship between rainfall and landslides by measuring near-surface water content variations over time. After appropriate calibration, the systems are revealed to be able to provide consistent soil moisture data. The sensors are rather cheap, easily relocatable and replaceable, which makes a remotely managed network relatively inexpensive and rapid to install.

Castelli et al. [3] report experimental data derived from laboratory and site tests, carried out with the aim of evaluating soil parameters for debris-flow run-out simulations. The experimental activities included triaxial tests in unsaturated conditions and geophysical investigations. Starting from a study of the geological framework and the historical background, this research focuses on the causes that triggered a recent debris-flow event in Sicily, Italy. The authors discuss possible strategies to determine the hydromechanical parameters that can be used for a numerical analysis of such phenomena and the role of matric suction, which reflects soil infiltration, in the investigation of rainfall-induced landslides.

Sitarenios and Casini [4] present a three-dimensional, limit equilibrium, slope stability solution for translational planar failure modes. The solution uses Bishop's average skeleton stress combined with the extended Mohr–Coulomb failure criterion. The performance of the proposed theoretical solution is evaluated by comparing its predictions with the results of a full-scale slope failure experiment carried out in Switzerland. The 3D solution can satisfactorily capture the experimentally observed behavior. Furthermore, its simplicity and very limited calculation costs allow for an easy evaluation of different scenarios with respect to slope geometries and soil parameters, which can be very useful in a landslide back-analysis.

Ventini et al. [5] discuss the assessment of the hydraulic and mechanical behavior of river embankments, by considering transient seepage under unsaturated soil conditions. This is a more accurate approach than assuming, as frequently occurs in routine engineering practice, steady-state seepage and either fully saturated or "dry" conditions. A comprehensive procedure for the definition of the key aspects of the problem is presented. The soil characterization in unsaturated conditions of a compacted mixture of sand and fines, typical of flood embankments of the main tributaries of the Po River in Italy, is reported. The laboratory results are used for parameter estimation, aiming to model the embankment performance under transient conditions and following a set of possible hydrometric peaks. The importance of carrying out such an analysis is highlighted in a comparison with steady-state seepage analyses performed at the maximum hydrometric height recorded, which frequently provide excessively conservative stability assessments. Safety factor variations over time are obtained by limit equilibrium analyses as the effect of river level fluctuations, providing useful information for further detailed experimental and numerical investigations on the river embankment response.

Sepe et al. [6] analyze the capillary barrier formation at the interface between shallow soil layers characterized by textural discontinuities. The investigation is focused on the effects of rainfall events in pyroclastic deposits of the Vesuvian Area, in Campania region, Italy, undergoing shallow instability phenomena caused by progressive soil saturation. Based on a detailed field survey, the failure surface location was identified at the interface between grey pumice and black scoriae layers, likely due to their contrasting hydraulic properties. One-dimensional infiltration into the stratified soil is numerically simulated, considering rainfall events of increasing intensity and duration. The changes in the degree of saturation and suction over time are used for the evaluation of stability condition variations. By analyzing the time history of the infiltration process, the authors highlight the capillary barrier water break-through as a consequence of the progressive suction equalization between the two layers, as well as the gravity-driven downward water flux over time. The simulated failure conditions due to a medium intensity, long duration rainfall event are consistent with the field observations.

Conflicts of Interest: The authors declare no conflict of interest.

References

1. Ponzoni, E.; Nocilla, A.; Jommi, C. Determination of Water Retention Properties of Silty Sands by Means of Combined Commercial Techniques. *Geosciences* **2021**, *11*, 315. [CrossRef]
2. Bovolenta, R.; Iacopino, A.; Passalacqua, R.; Federici, B. Field Measurements of Soil Water Content at Shallow Depths for Landslide Monitoring. *Geosciences* **2020**, *10*, 409. [CrossRef]
3. Castelli, F.; Lentini, V.; Venti, A.D. Evaluation of Unsaturated Soil Properties for a Debris-Flow Simulation. *Geosciences* **2021**, *11*, 64. [CrossRef]
4. Sitarenios, P.; Casini, F. The Hydromechanical Interplay in the Simplified Three-Dimensional Limit Equilibrium Analyses of Unsaturated Slope Stability. *Geosciences* **2021**, *11*, 73. [CrossRef]
5. Ventini, R.; Dodaro, E.; Gragnano, C.G.; Giretti, D.; Pirone, M. Experimental and Numerical Investigations of a River Embankment Model under Transient Seepage Conditions. *Geosciences* **2021**, *11*, 192. [CrossRef]
6. Sepe, C.; Calcaterra, D.; Cecconi, M.; Di Martire, D.; Pappalardo, L.; Scarfone, R.; Vitale, E.; Russo, G. Capillary Barriers during Rainfall Events in Pyroclastic Deposits of the Vesuvian Area. *Geosciences* **2021**, *11*, 274. [CrossRef]

geosciences

MDPI

Article

Determination of Water Retention Properties of Silty Sands by Means of Combined Commercial Techniques

Elisa Ponzoni [1], Alessandra Nocilla [2,*] and Cristina Jommi [3,4]

[1] Rina S.p.A., Via Antonio Cecchi, 6, 16129 Genova, Italy; elisa.ponzoni@rina.org
[2] Dipartimento di Ingegneria Civile Ambiente Territorio Architettura e Matematica, University of Brescia, Via Branze, 43, 25123 Brescia, Italy
[3] GSE Laboratory, Delft University of Technology, Mekelweg 5, 2628 CD Delft, The Netherlands; c.jommi@tudelft.nl or cristina.jommi@polimi.it
[4] Dipartimento di Ingegneria Civile e Ambientale, Politecnico di Milano, Piazza Leonardo da Vinci, 32, 20133 Milano, Italy
* Correspondence: alessandra.nocilla@unibs.it

Citation: Ponzoni, E.; Nocilla, A.; Jommi, C. Determination of Water Retention Properties of Silty Sands by Means of Combined Commercial Techniques. *Geosciences* **2021**, *11*, 315. https://doi.org/10.3390/geosciences11080315

Academic Editors: Roberto Vassallo, Luca Comegna, Roberto Valentino and Jesus Martinez-Frias

Received: 7 June 2021
Accepted: 20 July 2021
Published: 27 July 2021

Publisher's Note: MDPI stays neutral with regard to jurisdictional claims in published maps and institutional affiliations.

Abstract: A recent increase in frequency and severity of exceptional climatic events is of concern for the stability of natural and artificial slopes. These undergo continuous evaporation and infiltration cycles, which change the suction distribution and trigger shrinkage, swelling, cracking, and surficial erosion, overall decreasing the soil strength. To assess the impact of these climatic stresses, the determination of water retention properties is a priority. Although advanced techniques have been proposed in the last few decades to this end, simpler commercially available techniques allow collecting data for a larger number of samples in a shorter time, thus enabling a basic description of the water retention properties for a larger database of soils. Data on two silty sands, coming from very different climatic environments in Europe, were collected with a combination of two simple commercial devices, and the results were modelled with a van Genuchten's law. The fitted parameters were found to correlate well with the amount of fines, irrespective of the different origin and composition of the two soils. Eventually, the limitation of the approach is discussed based on the results of cyclic drying–wetting tests.

Keywords: water retention; suction; silty sands; commercial experimental techniques

1. Introduction

Slope stability analyses involve the identification of the different mechanisms that might trigger failure no matter whether an artificial or a natural slope is involved. A recent increase in the severity of climatic events suggests that the influence of evaporation and infiltration on the mechanical behaviour of soils must be considered carefully among other factors. Two recent case histories, very different from each other, promoted the investigation presented in the following. The first one refers to the stability of regional dykes in The Netherlands, which are experiencing droughts due to substantial increase in summer temperatures in the recent years [1]. The second case refers to natural cliffs made of carbonate sand at Agrigento in Italy, where local collapses have been observed following heavy rainfalls [2–5]. Both cases involve large areas, for which management decisions have to be taken based on a relatively small amount of data, as advanced lengthy laboratory investigation cannot be afforded.

The hydromechanical response of dykes under unsaturated conditions has been tackled by many researchers, who highlighted the role of unsaturated conditions on their response. Among others, Schmertmann (2006) and Calabresi et al. (2013) [6,7] highlighted that partial saturation is largely prevailing in silty dykes. As a consequence, the current guidelines, which suggest assessing slope stability assuming a steady state pore pressure distribution following the design high-water event, are largely over conservative. As pointed out by Jommi and Della Vecchia (2013) [8], the steady state assumption leads to an overestimation of the dimensions of the embankment, hence of the amount of soil, needed to guarantee a sufficient factor of safety during the working life of the structure. As is well-known from the literature, modelling time dependent unsaturated flow better allows assessing the change in the pore water pressure and in the soil shear strength [9].

Things are not less complex for natural slopes. Of interest for the following, it is worth noting that Ercoli et al. (2015) [3] suggested that monitoring the change in water content in the sand is fundamental to appreciate its main role in the retrogression collapse mechanisms of the Valle dei Templi cliffs at Agrigento.

The water retention properties are fundamental in this respect, not only because they govern the hydraulic conductivity and seepage, but eventually for their role in governing changes in stiffness and strength of the soil. Many factors affect the water retention, including the void ratio, the activity of the clay fraction, the fabric of the soil [10–18]. Among them, the particle size distribution may become the dominant one for non-plastic mixtures of silty sands [19,20].

Although advanced techniques have been proposed in the last few decades to study the water retention behaviour of soils, simpler commercially available techniques allow collecting data for a larger number of samples in a shorter time, thus enabling a basic description of the water retention properties for a larger database of soils. Following this consideration, two non-plastic soils were investigated over drying and wetting combining the data from two commercially available devices: the Hyprop® and the Dew Point Potentiometer WP4C®. The results are modelled with a simple van Genuchten's model (1980) [21], in an attempt to derive correlations to be used as a first approximation in the assessment of natural and artificial slopes.

2. Materials and Methods

Two different soils were investigated: a sandy silt from a dyke at the Leendert de Boerspolder (South Holland, The Netherlands) and a silty carbonate sand from a natural cliff in the "Valle dei Templi" at Agrigento (Sicily, Italy). Relevant information on the samples tested are collected in Table 1. The dyke silty material goes from a silty sand with traces of gravel and clay to a clayey silt with traces of sand. The void ratio, e, of the samples investigated is almost constant, to better highlight the influence of grain size distribution on the water retention properties. Both natural and reconstituted samples were investigated to assess the influence of fabric on the response. The material from the Valle dei Templi is a carbonate fine and medium sand with a clay fraction up to 10% and a silt content slightly less than 20%, characterised by a quite high uniformity coefficient, U, of about 5 [22]. For the Agrigento sand the values reported are the average of the three samples. For both materials the amount of clay is below 8%.

Table 1. Samples investigated for water retention. In the table, r stands for reconstituted.

Sample	Origin	Gs	w	Sr	e	Fines%	Sand%	Gravel%
L1	Leendert	2.6	0.27	0.93	0.75	32	63	5
L2	de Boer-	2.64	0.25	0.88	0.73	57	43	-
L3	spolder	2.58	0.25r	0.91r	0.73	45	54	1
A1			0.21r	0.77r	0.73			
A2	Agrigento	2.7	0.19r	0.76r	0.67	≈25	≈73	≈2
A3			0.19r	0.72r	0.74			

The water retention behaviour was investigated by combining data from two commercial apparatuses (Figure 1) available at TU Delft (NL), namely the Hyprop® and the Dew Point Potentiometer WP4C®. The HYPROP® device (HYdraulic PROPerty analyser, constructed in Munich, Germany) was developed by UMS to determine the water retention curve and the unsaturated conductivity over drying from the top surface of the sample at room temperature and relative humidity. Over the testing period the total mass of the sample is measured, as well as the temperature at the bottom of the sample. The sensor placed at the bottom of the sample can measure temperature between -30 and $70\ ^\circ$C, with a tolerance of $\pm0.2\ ^\circ$C in the range of -10 to $30\ ^\circ$C. Two tensiometers, having a porous stone with a high entry value (880 kPa), allow direct measurement of suction at two different heights, with an accuracy of 0.15 kPa. The range of suctions typically covered by the tensiometers goes from zero to about 100 kPa, depending on the saturation achieved by the suction measurement system before the start of the test. Actually, although the porous stone in contact with the soil guarantees no cavitation of the tip, the longer the tests the more likely is cavitation of the measuring system behind the tip. Water mass changes are measured by a scale (accuracy of 0.1 g) at the bottom of the system, which allows estimating the average water ratio of the sample at any time. The water content at the end of each test was double checked by oven drying.

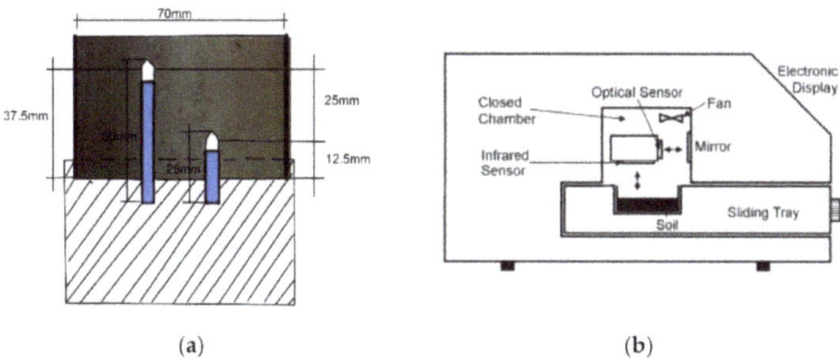

Figure 1. Geometry and cross section of (**a**) the Hyprop device edited from the User Manual Hyprop, 2012 [23], and (**b**) The Decagon dew point potentiometer WP4C® Cross Section [24].

The Dew Point Potentiometer, constructed in Munich, Germany, uses the chilled mirror dew point technique to measure the water potential of the soil, through the vapour pressure of the air in equilibrium with the sample in a sealed measurement chamber. Although the measurement range theoretically goes from 0 to 300 MPa, the accuracy of the device below 1MPa is extremely low, which does not guarantee the reliability of the measurement. From 10 to 300 MPa the accuracy is $\pm1\%$.

Although both techniques have been designed to investigate the water retention properties over drying, an attempt was made to conduct the tests in order to investigate also the response of the samples over wetting. However, it is worth noting that the data obtained from the Dew Point Potentiometer after wetting the samples cannot be interpreted as lying on a main wetting curve, as equilibrium is achieved in the chamber due to an evaporation process [25]. In the Hyprop, wetting stages were controlled by protecting the samples from evaporation after adding a controlled amount of water on the top. The recorded data over drying and wetting will be commented later in Section 4.

3. Results

The mass of water contained in a soil samples allows determining directly the water content of the soil, w

$$w = \frac{M_w}{M_s} \tag{1}$$

where M_w and M_s are the mass of water and solid particles, respectively.

Following the discussion by Romero and Vaunat (2000) [26], water ratio, e_w, is used instead of water content in the presentation of the results, as it better normalises the results for soils having different specific gravity, G_s:

$$e_w = G_s \cdot w \tag{2}$$

For soils having low deformability like those tested, volumetric strains over drying in the range of suctions investigated in the Hyprop can be disregarded as a first approximation, thereby allowing a sufficiently reliable estimate of the current average degree of saturation of the sample assuming constant void ratio over the test ($e = e_0$).

$$S_r = \frac{e_w}{e_0} \tag{3}$$

Typical data for the two materials over main drying are reported in Figure 2. During the Hyprop test water tension readings (suction) are taken continuously by the two tensiometers. The suction in the upper part of the sample is measured by the tensiometers having the longer shaft and that in the lower part of the sample by the tensiometers mounted on the shorter shaft. Both shafts are schematized in Figure 1a. By contrast, the discrete measurements taken with the WP4C are represented by symbols in the figures.

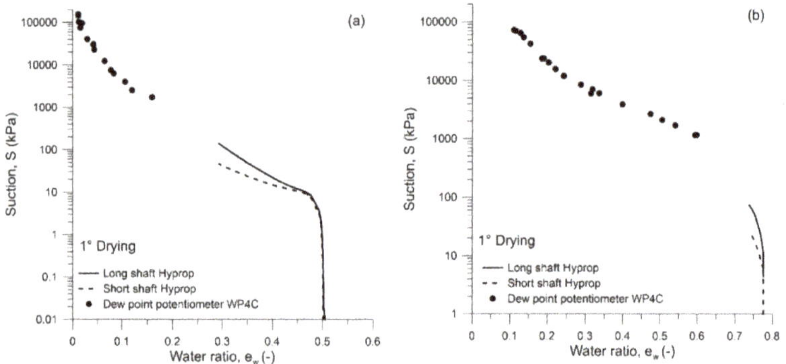

Figure 2. Water retention data of (**a**) Agrigento sand and (**b**) Leendert de Boerspolder silt.

Figure 2 highlights that combining the two techniques allows covering the entire range of suction expected for these soils, although the suction range from about 100 kPa to 1 MPa is not covered by any of the two techniques. Nonetheless, the lower suction and higher suction ranges are accurately described, which allows including the entire suction range in a model.

It is worth noting that the suction at the bottom and at the top of the samples are different, as the measurements are taken over a time-dependent process characterised by vertical flow. Moreover, the two measurement are related to a water ratio, which actually measures the average water mass on the entire sample coming from a non-uniform distribution over sample height. Nevertheless, for rather pervious almost undeformable soils the average suction over height and the average water ratio over the volume of the sample are well correlated to each other because their gradients are not high. This is not the case for more impervious and deformable clay samples, which present steep gradients of water content and suction over height due to the preferential flow direction in time [27]. Assuming that the average values are representative of the "material" behaviour, the water retention data can be elaborated to derive a degree of saturation–suction relationship for each sample, as shown in Figure 3.

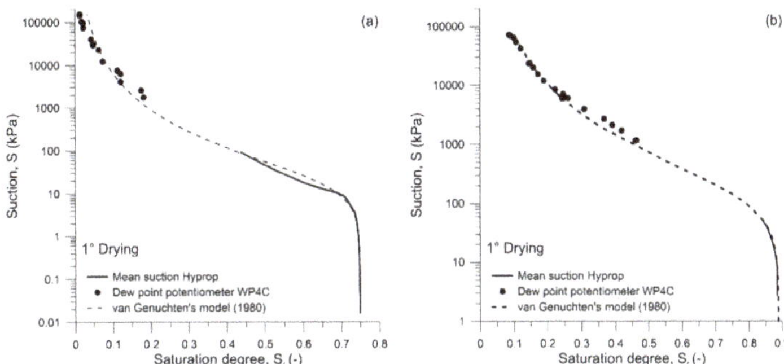

Figure 3. Elaboration of water retention data of (**a**) Agrigento sand and (**b**) Leendert de Boerspolder silt.

The elaborated data are interpolated with a simple van Genuchten's model (1980):

$$S_r = \left(1 + \left(\frac{S}{P}\right)^{1/1-\lambda}\right)^{-\lambda} \cdot (S_{rmax} - S_{rres}) + S_{rres} \tag{4}$$

drawn as a dotted line in the figure. In the previous expression, S is the suction, S_r is the degree of saturation, S_{rmax} is the maximum attainable degree of saturation over wetting at the atmospheric pressure, and S_{rres} is the residual degree of saturation at high suctions. The curve depends on two parameters, p linked to the air entry value and λ, ruling the average slope of the curve. For the samples investigated, the residual saturation is almost null, while the maximum attainable saturation at the atmospheric pressure corresponds with the air occlusion value, as the samples did not undergo saturation under back pressure. Strictly speaking, the initial part of the plotted drying curve for Agrigento's soil (until the air entry value) is not the "main drying" because the soil sample did not start from full saturation. However, it is equally representative of the response of the soil. The parameters of the fitted van Genuchten's curve for the different samples are summarised in Table 2, where p has been normalised to the atmospheric pressure.

Table 2. Parameters of the van Genuchten's model for the samples investigated.

Sample	Void Ratio	Initial Water Ratio	Maximum Saturation Degree	Fine Content	p/p_{atm}	λ
	e	e_{w0}	S_{rmax}	FC	-	-
L1	0.75	0.7	0.93	0.32	0.5	0.29
L2	0.73	0.66	0.88	0.57	1.1	0.32
L3	0.73	0.65	0.91	0.45	0.8	0.29
A1, A2, A3	≈0.73	≈0.54	≈0.77	0.25	0.3	0.28

The calibrated parameters are very well correlated with the fine content, including the silt and the clay fraction, of each sample. The correlations are reported in Figure 4, and the curves fitting the available experimental data, starting from the experimental S_{rmax}, are depicted in Figure 5.

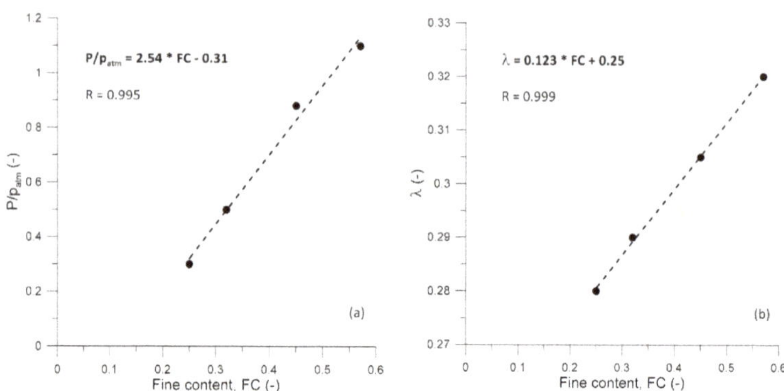

Figure 4. Fitting curves of the water retention model parameters as a function of the fine content, FC: (a) p/p_{atm}, and (b) λ.

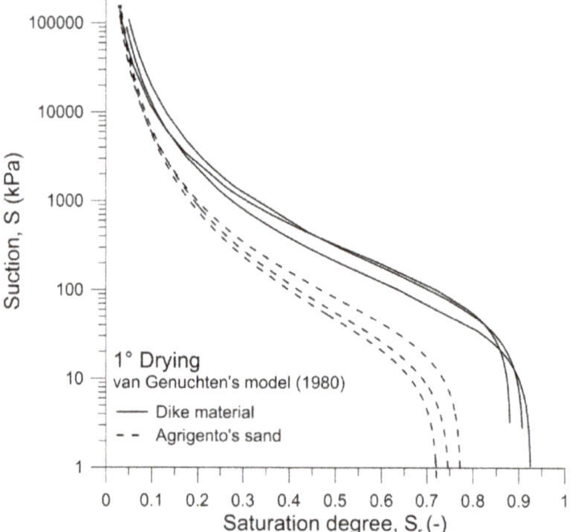

Figure 5. Fitting curves of the samples investigated.

Although the amount of data is limited, the correlations seem to suggest few conclusions: (i) the fabric does not play a relevant role for these samples, as both natural and reconstituted samples follow the same trend; (ii) as expected, the air entry value increases with the amount of fine, irrespective of the mineralogical composition of the two soils, which in both case is non-active; (iii) the slope of the water retention curves also depends on the amount of fines, as it rules the coefficient of uniformity. Given the very different origin of the two soils, it is expected that the conclusions may be valid for non-active silty sands in general, although validation is still needed.

4. Time-Dependent Data and Hysteresis

Although the Hyprop system is designed to determine the water retention curve over drying at atmospheric boundary conditions, an attempt was made to investigate the soil water retention branch. As suggested by Schindler et al. (2015) [28], water can be added to the surface of the Hyprop sample simulating a precipitation event in order to quantify hysteresis due to cycles of drying and wetting.

The data recorded over time over a number of drying and wetting cycles on a sample of the dyke silty material are reported in Figure 6, as a function of the current degree of saturation. The shape of the curves obtained by averaging the negative pore pressure recorded by the two tensiometers clearly highlight the transient response of the system, especially over the first cycles, and hardly allows determining the separate main drying and main wetting branches of the water retention domain. Actually, whenever forced infiltration is stopped, the upper part of the samples undergoes a new drying stage due to evaporation from the top, while the bottom part of the sample is still wetting due to previous infiltration.

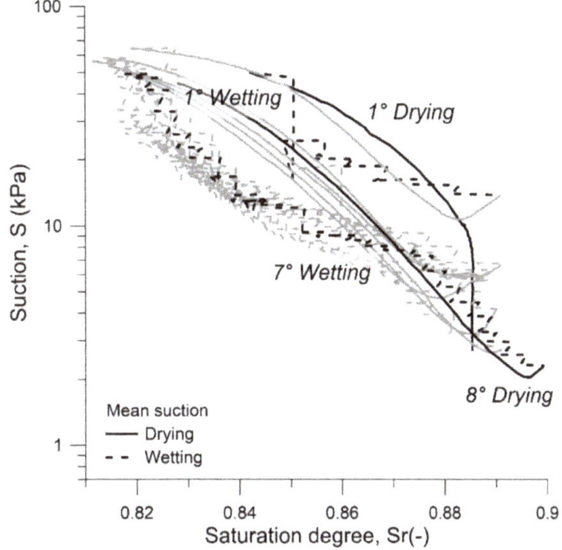

Figure 6. Suction measurements during drying–wetting cycles in Hyprop on a sample of the dyke silty material as a function of the average degree of saturation of the sample.

Therefore, it was decided to adopt a slightly different strategy, which could help in the interpretation of the data. After adding water on the top of the sample, an impervious plastic cover was placed, and kept until the tensiometers measurement reached an almost asymptotic value of suction, indicating equalisation, before letting the sample evaporate again. An example of the entire data set logged during evaporation and infiltration cycles is provided in Figure 7 for another sample of the dyke silty sand. In Figure 7a the inner and outer temperature are plotted against time. The term "inner" indicates the temperature recorded at the base of the sample from the sensor placed in the Hyprop unit, whereas "outer" indicates the ambient temperature of the laboratory. Figure 7b shows the mass change of the sample over time. Its rate is plotted as a function of the two variables ruling the exchanges, temperature and relative humidity, in Figure 7c.

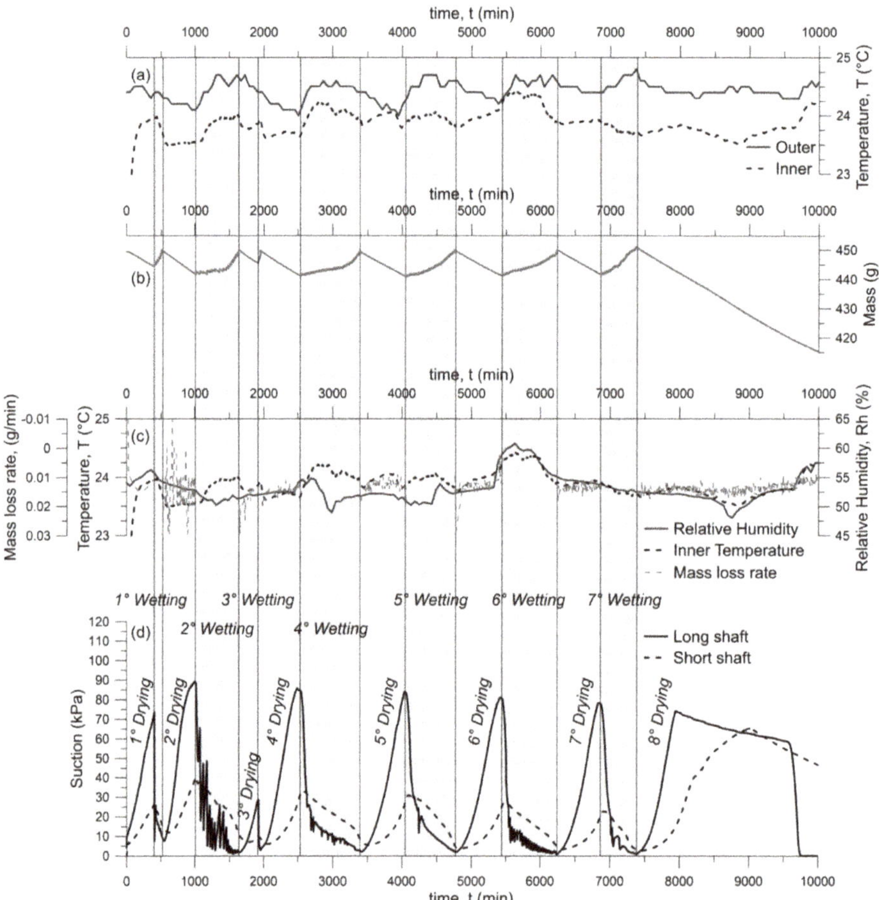

Figure 7. Data logged during drying and wetting cycles on a sample of the dyke silty material: (**a**) inner and outer temperature, (**b**) temperature, Rh and mass loss rate, (**c**) net sample mass and (**d**) suction readings as a function of time.

Of relevance for the present discussion, the suction measured on the upper part of the soil sample (long shaft) and in the lower one (short shaft) are reported in Figure 7d. The measurements clearly show the substantial difference between the suction at the two heights over time during drying, when equalisation was not forced. The maximum suction over drying is reached in the lower part of the sample shortly after the peak value at the top. If equalisation is forced, as it was done over the wetting stages here, the two tensiometers reach almost the same value (with a difference due to the gravitational potential, which can be disregarded) and allow for the determination of a clear value for the suction in equilibrium with a given water content, to be used in a water retention model. The decrease in the measured suction on the long shaft tensiometer during the 8th drying stage is due to cavitation of the instrument.

The latter data were elaborated to identify the wetting branch of the water retention curve of the sample, which is plotted in Figure 8, together with the drying data, which were already used as example in Figure 2b. The hysteretic cycles could be well identified, as shown in Figure 8b, with the procedure adopted, with the scanning paths that tend to overlap to each other after the first cycle, as expected.

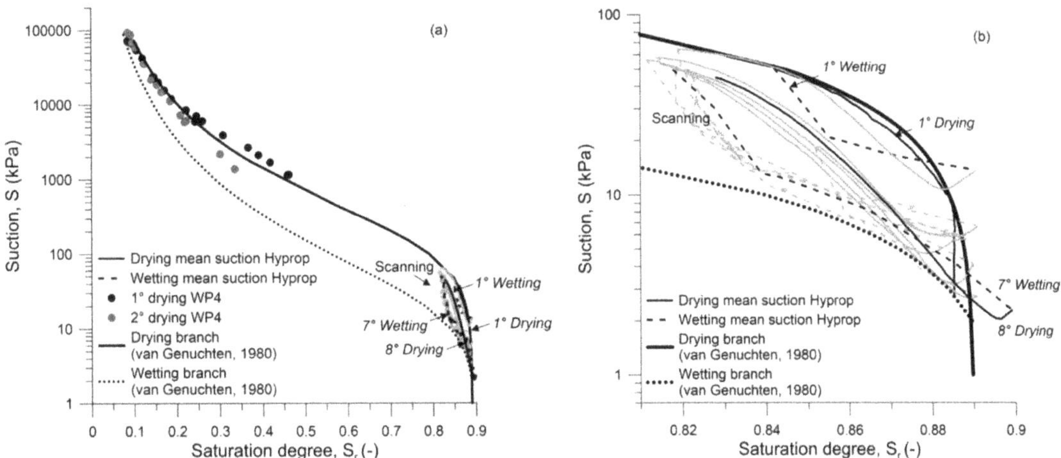

Figure 8. Drying and wetting branches of the water retention domain, as elaborated from the Hyprop data in Figure 7 on a sample of dyke silty material: (**a**) full suction range, and (**b**) 0 < S < 100 kPa.

5. Conclusions

Soil water retention properties over drying were investigated on a sandy silt from a dyke at the Leendert de Boerspolder (NL) and for a carbonate silty sand from the Valle dei Templi in Agrigento (IT), with a combination of data obtained from two commercially available devices, namely the Hyprop® and the Dew Point Potentiometer WP4C®. The scope of the work was to investigate the possibility for a rather cheap characterisation of the retention properties of soils of concern for the assessment of climate impact on the stability of slopes, even in very different contexts.

The results demonstrate the feasibility of the approach, and provided careful evaluation and elaboration of the results. Classical water retention models are conceived to describe steady equilibrium states, while in the popular devices chosen for the investigation, data are continuously recorded over time during time-dependent dynamic processes. Using the average over a given time interval of the suction measurements taken at two different heights in the sample may work for the first drying stage of non-active silty sands and sandy silts. However, to obtain reliable information on the hysteretic response over the drying–wetting cycles expected on site, the standard procedures must be amended.

An advantage of the technique is that it reproduces evaporation and infiltration naturally occurring in the soil at the environmental conditions of the laboratory, which can be controlled if necessary. However, the way in which the Hyprop device was designed implies measuring suction during a transient process, which might even be non-monotonic over the sample height in case wetting stages were added during the test. With minor changes to the set up and the test protocol of the Hyprop device, data at suction equalisation can be obtained, creating the possibility to investigate both the drying and the wetting branches of the water retention domain of non-active soils. Soils experiencing relevant volume changes over drying and wetting are much less suitable for testing with the technique and the coupled hydromechanical response of the soil requires careful investigation.

Author Contributions: Conceptualization, A.N.; Investigation, E.P.; Methodology, C.J.; Writing—original draft, A.N., E.P.; Writing—review & editing, A.N., C.J. All authors have read and agreed to the published version of the manuscript.

Funding: The financial support on the experimental equipment of the perspective program BioGeoCivil (grant 11344) and CEAMaS (Civil Engineering Applications for Marine Sediments), supported by the European Regional Development Funding through INTERREG IV B, is gratefully acknowledged.

Institutional Review Board Statement: Not applicable.

Informed Consent Statement: Not applicable.

Data Availability Statement: Not applicable.

Acknowledgments: The authors acknowledge Margherita Zimbardo of the University of Palermo and the "Ente Parco Archeologico della Valle dei Templi" at Agrigento (Sicily, Italy) for providing the Agrigento sand specimens.

Conflicts of Interest: The authors declare no conflict of interest.

References

1. Chao, C.Y.; Bakker, M.; Jommi, C. *Calibration of a Simple 1D Model for the Hydraulic Response of Regional Dykes in The Netherlands*; Cardoso, R., Jommi, C., Romero, E., Eds.; E-UNSAT: Amsterdam, The Netherlands, 2020; Volume 195, pp. 1–6.
2. Cotecchia, V. Geotechnical degradation of the archaelogical site of Agrigento. In *Arrigo Croce Memorial Symposium*; Balkema: Kalamazoo, MI, USA, 1996.
3. Ercoli, L.; Zimbardo, M.; Nocilla, N.; Nocilla, A.; Ponzoni, E. Evaluation of cliff recession in the Valle dei Templi in Agrigento (Sicily). *Eng. Geol.* **2015**, *192*, 129–138. [CrossRef]
4. Zimbardo, M.; Ercoli, L.; Megna, B. The open metastable structure of a collapsible sand: Fabric and bonding. *Bull. Int. Assoc. Eng. Geol.* **2015**, *75*, 125–139. [CrossRef]
5. Nocilla, A.; Grossi, G.; Ponzoni, E. Influence of the soil-atmosphere interaction on the collapse of sand in the Valle dei Templi in Agrigento. In Proceedings of the VI Italian Conference of Researchers in Geotechnical Engineering CNRIG 2016 "Geotechnical Engineering in multidisciplinary research: From microscale to regional scale", Bologna, Italy, 22–23 September 2016; Volume 158, pp. 386–391.
6. Schmertmann, J.H. Estimating Slope Stability Reduction due to Rain Infiltration Mounding. *J. Geotech. Geoenviron. Eng.* **2006**, *132*, 1219–1228. [CrossRef]
7. Calabresi, G.; Colleselli, F.; Danese, D.; Giani, G.; Mancuso, C.; Montrasio, L.; Nocilla, A.; Pagano, L.; Reali, E.; Sciotti, A. Research study of the hydraulic behaviour of the Po River embankments. *Can. Geotech. J.* **2013**, *50*, 947–960. [CrossRef]
8. Jommi, C.; Della Vecchia, G. Geomechanical analysis of river embankments. In *Mechanics of Unsaturated, Geomaterials*; Laloui, L., Ed.; John Wiley & Sons: Hoboken, NJ, USA, 2013; pp. 353–373.
9. Fredlund, D.G.; Rahardjo, H. *Soil Mechanics for Unsaturated Soils*; John Wiley & Sons: Hoboken, NJ, USA, 1993. [CrossRef]
10. Vanapalli, S.K.; Fredlund, D.G.; Pufahl, D.E. The influence of soil structure and stress history on the soil–water characteristics of a compacted till. *Geotechnique* **1999**, *49*, 143–159. [CrossRef]
11. Fredlund, M.D.; Fredlund, D.G.; Wilson, G.W. Estimation of volume change functions for unsaturated soils. In *Unsaturated Soils for Asia*; Rahardjo, H., Toll, D.G., Leong, E.C., Eds.; CRC Press: Boca Raton, FL, USA, 2000; pp. 663–668.
12. Kawai, K.; Kato, S.; Karube, D. The model of water retention curve considering effects of void ratio. In *Unsaturated Soil for Asia*; Rahardjo, H., Toll, D.G., Leong, E.C., Eds.; CRC Press: Boca Raton, FL, USA, 2000; pp. 329–334.
13. Cafaro, F. Metastable states of silty clays during drying. *Can. Geotech. J.* **2002**, *39*, 992–999. [CrossRef]
14. Gallipoli, D.; Wheeler, S.; Karstunen, M. Modelling the variation of degree of saturation in a deformable unsaturated soil. *Geotechnique* **2003**, *53*, 105–112. [CrossRef]
15. Tarantino, A.; Tombolato, S. Coupling of hydraulic and mechanical behaviour in unsaturated compacted clay. *Geotechnique* **2005**, *55*, 307–317. [CrossRef]
16. Romero, E.; DELLA Vecchia, G.; Jommi, C. An insight into the water retention properties of compacted clayey soils. *Geotechnique* **2011**, *61*, 313–328. [CrossRef]
17. Casini, F.; Vaunat, J.; Romero, E.; Desideri, A. Consequences on water retention properties of double-porosity features in a compacted silt. *Acta Geotech.* **2012**, *7*, 139–150. [CrossRef]
18. Gomez, R.; Romero, E.; Della Vecchia, G.; Jommi, C.; Suriol, J.; Lloret, A. Water Retention Properties of a Compacted Clayey Silt Including Void Ratio Dependency and Microstructural Features. In *Unsaturated Soils: Research and Applications*; Mancuso, C., Jommi, C., D'Onza, F., Eds.; Springer: Berlin/Heidelberg, Germany, 2012; pp. 153–158. [CrossRef]
19. Cafaro, F.; Hoffmann, C.; Cotecchia, F.; Buscemi, A.; Bottiglieri, O.; Tarantino, A. Modellazione del comportamento idraulico di terreni parzialmente saturi a grana media e grossa. *Riv. Ital. Geotec.* **2008**, *3*, 54–72.
20. Wang, M.; Pande, G.; Pietruszczak, S.; Zeng, Z. Determination of strain-dependent soil water retention characteristics from gradation curve. *J. Rock Mech. Geotech. Eng.* **2020**, *12*, 1356–1360. [CrossRef]
21. Van Genuchten, M.T. A Closed-form Equation for Predicting the Hydraulic Conductivity of Unsaturated Soils. *Soil Sci. Soc. Am. J.* **1980**, *44*, 892–898. [CrossRef]
22. Nocilla, N.; Ercoli, L.; Zimbardo, M.; Nocilla, A.; Meli, P.; Grado, G.; Parello, G.; Presti, G. Unsaturated sand in the stability of the cuesta of the Temple of Hera (Agrigento). In *Geotechnical Engineering for the Preservation of Monuments and Historic Sites*; Bilotta, E., Flora, A., Lirer, S., Viggiani, C., Eds.; Taylor and Francis: London, UK, 2013; Volume 1, pp. 603–612. [CrossRef]
23. HYPROP© UMS GMBH MUNICH. User Manual Hyprop. Available online: www.ums-muc.de (accessed on 1 June 2015).

24. Bulut, R.; Hineidi, S.M.; Bailey, B. Suction measurements—Filter paper and chilled mirror psychrometer. In Proceedings of the Texas Section American Society of Civil Engineers, Fall Meeting, Waco, TX, USA, 2–5 October 2002.
25. Cardoso, R.; Romero, E.; Lima, A.; Ferrari, A. A Comparative Study of Soil Suction Measurement Using Two Different High-Range Psychrometers. In *Experimental Unsaturated Soil Mechanics*; Schanz, T., Ed.; Springer: Berlin/Heidelberg, Germany, 2007; Volume 112, pp. 79–93. [CrossRef]
26. Romero, E.; Vaunat, J. Retention curves of deformable clays. In *Experimental Evidence and Theoretical Approaches in Unsaturated Soils*; Tarantino, A., Mancuso, C., Eds.; CRC Press: Boca Raton, FL, USA, 2000; pp. 91–106.
27. Tollenaar, R.N.; van Paassen, L.; Jommi, C. Small-scale evaporation tests on clay: Influence of drying rate on clayey soil layer. *Can. Geotech. J.* **2018**, *55*, 437–445. [CrossRef]
28. Schindler, U.; von Unold, G.; Müller, L. Laboratory measurement of soil hydraulic functions in a cycle of drying and rewetting. *Int. J. Emerg. Technol. Adv. Eng.* **2015**, *5*, 282–286.

Article

Field Measurements of Soil Water Content at Shallow Depths for Landslide Monitoring

Rossella Bovolenta ***, Alessandro Iacopino, Roberto Passalacqua and Bianca Federici**

Department of Civil, Chemical and Geotechnical Engineering, University of Genoa, 16145 Genoa, Italy;
a_iacopino.a@libero.it (A.I.); roberto.passalacqua@unige.it (R.P.); bianca.federici@unige.it (B.F.)
* Correspondence: rossella.bovolenta@unige.it; Tel.: +39-010-335-2505

Received: 9 August 2020; Accepted: 10 October 2020; Published: 13 October 2020

Abstract: Monitoring changes in soil saturation is important for slope stability analyses. Soil moisture capacitive sensors have recently been developed; their response time is extremely fast, they require little maintenance, and they are relatively inexpensive. The use of low-cost sensors in landslide areas can allow the monitoring of large territories, but appropriate calibration is required. Installation in the field and the setting up of the monitoring network also require attention. In the ALCOTRA AD-VITAM project, the University of Genoa is involved in the development of a system, called LAMP, for the monitoring, analysis and forecasting of slides triggered by rainfalls. Multiple installations (along vertical alignments) of WaterScout sensors are placed in the nodes of the monitoring network. They provide real-time water content profiles in the shallow layers (typically in the upper meter) of a slope. With particular reference to these measurements, the present paper describes the reliability analysis of the instruments, the operations related to the sensor calibration and the installation phases for the monitoring networks. Finally, some of the data coming from a node, belonging to one of the five monitoring networks, are reported.

Keywords: landslide; soil slide; LAMP; soil water content; soil moisture; monitoring; calibration; installation; rainfall

---

1. Introduction

Rainfall affects the stability of natural slopes, excavation fronts and loose material works, such as embankments. Rainwater infiltration increases the degree of saturation and pore-water pressure, which results in reduced shear strength.

In evaluating the landslide susceptibility of a territory, the local hydrological and soil chemo-mechanical properties/conditions and their variability (both in space and in time) are particularly important.

In general, the unsaturated zone can be monitored by measuring the suction and/or water content. The measurement of suction may be rather problematic [1], while the measurement of (volumetric) water content with sensors, based on the measurement of the soil's dielectric permittivity, appears to be more feasible. Capacitive type sensors respond "immediately" to changes in soil moisture, require moderate maintenance, and are relatively inexpensive, but a more laborious calibration is required [2,3].

The use of low-cost probes allows the monitoring of large portions of territory, providing continuous and real-time data at various depths in order to define soil moisture profiles. By combining these measurements with rainfall, piezometers and displacement data, recorded as continuously as possible, hydrological and geotechnical analyses can be conducted properly.

Since there is no simple relationship between rainfall and the hydraulic conditions of the slopes at the depths where the sliding surfaces develop, the measurement of the surface moisture conditions of the soil is interesting for the study of landslide phenomena [4,5] in combination with physically based analysis models. The evaluation of the water content allows the analysis of the soil's behaviour from a hydrological point of view and the study of the mechanical performance under partial-saturation conditions. Experiments have shown how the degree of saturation affects shear strength, compressibility and stiffness in the range of small strains [6,7].

The measurement of water content may also be useful for the study of vegetation cover; it is well known that vegetation modifies the soil's strength [8–10], as well as the hydrological balance [11]. Therefore, the measurement of soil moisture is very important for the analysis of natural and artificial slopes in partially saturated soils [12–14].

As part of the project AD-VITAM (Analysis of the Vulnerability of the Mediterranean Alpine Territories to natural risks), a cross-border cooperative project financed within the Interreg V-A France–Italie, ALCOTRA 2014–2020, the University of Genoa has developed a system, called LAMP (LAndslide Monitoring and Predicting), for the analysis and forecasting of landslides (more specifically, soil slides) triggered by rainfalls [15,16]. LAMP comprises an Integrated Hydrological–Geotechnical model [17], fed by a Wireless Sensor Network (WSN), including capacitive sensors for soil moisture measurements.

The AD-VITAM project aims to improve the resilience of territories in the face of the risk of landslides induced by rainfalls, bringing together the skills of research groups with those of the involved local technicians. Within the project, five sites have been chosen in which to apply the LAMP system; four of them are located in the province of Imperia (Liguria, Italy), and the fifth is in the PACA Région, Provençal area (France). All of these sites have slope-instability phenomena, triggered by rainfall. When equipping the sites with the LAMP system, practical support is offered to the local authorities, who oversee the territorial planning and public safety.

In the present paper, particular attention is devoted to the soil moisture sensors, the monitoring network's installation phases, analysis of the instruments' reliability, and operations related to the calibration of the sensors.

2. Materials and Methods

LAMP is based on an Integrated Hydrological–Geotechnical model, hereinafter referred to as an IHG model, and a low cost, self-sufficient, remotely manageable monitoring network that feeds the IHG model, allowing quasi-real-time analyses.

The following brief description of the LAMP system anticipates a focus on soil water content sensors. This research not only highlights the behaviour of such sensors in relation to certain rain events but also provides fundamental indications for following research aimed at improving the response of the instruments, the interpretation of the obtained data and the installation procedures. The main research activities, including the field tests, installation on site, sensor calibration and feeding of the IHG model are illustrated in Figure 1 in chronological order.

2.1. LAMP—LAndslide Monitoring and Predicting

LAMP makes it possible to assess susceptibility to landslides (specifically, debris and earth slides) in the event of measured or expected precipitations, establishing a cause–effect link between rainfalls and landslides. It is based on the IHG model [18], which is physically based and developed in a Geographic Information System (GIS) environment. The IHG model is designed to analyse areas of a few square kilometres in a short period of time, so that the response of the site under examination can be assessed in quasi-real time according to the monitoring data (the rain history, temperature and water content).

The characterization of a site requires knowledge of the stratigraphy, physical and strength parameters, permeability and piezometer measurements.

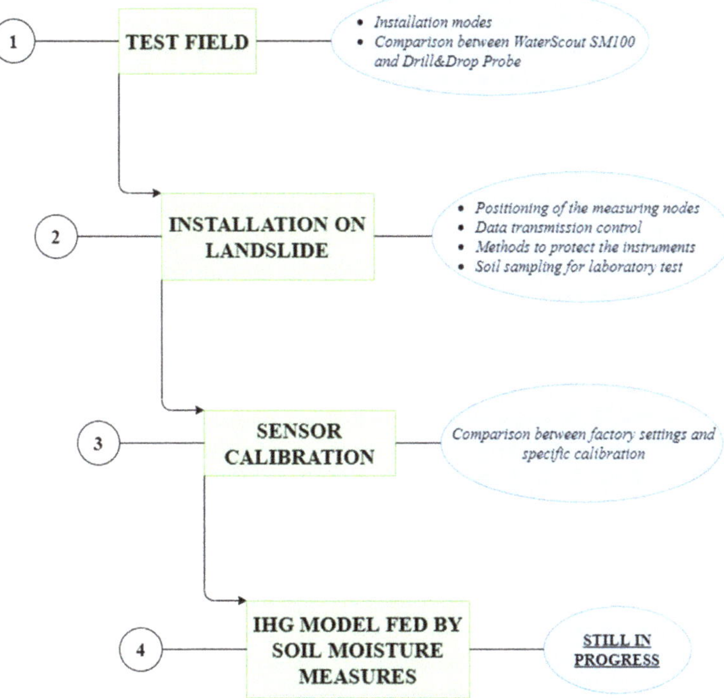

Figure 1. Chronological chart of the research activities.

The modelling is three-dimensional, both geometrically and in terms of geotechnical and hydrological parameters; these, in particular, are spatially distributed by appropriate methods of interpolation/extrapolation on data from surveys and in situ monitoring [19]. For any studied site, a 3D model is implemented in the GIS environment. Its surface is discretized in pixels, typically 5 m × 5 m, i.e., the resolution of the Digital Terrain Model from which the upper surface of the model is generally obtained. The model is bounded below by a surface, which usually corresponds to the bedrock or a stable layer. Additional internal surfaces can be inserted to identify several soil layers.

The groundwater table is also modelled, and can fluctuate according to weather conditions; its fluctuations are evaluated by hydrological balance. In fact, the landslide volume is divided into a series of three-dimensional elements hydraulically considered to be like underground reservoirs. The model is fed by the rainfall, which can vary in both space and time [20]. Each discretization cell is studied as a reservoir, where the infiltrated rainwater enters to exit by downhill seepage and evapotranspiration.

For a given rainfall event (Figure 2), the amount of infiltrating water is calculated using the Modified CN Method (the Soil Conservation Service Curve Number method was first published in 1956 [21]). For this estimation, it is fundamental to know the land cover and soil moisture conditions three days before the analysed rainfall event. In fact, the hydro-mechanical and meteorological conditions preceding a rainfall event are very important [22,23] in the slope response.

The stability analysis is based on the global limit equilibrium method applied to each three-dimensional element. In analogy to [24] and other models in the literature [25–28], it is assumed that the failure surface is parallel to the ground plane. In the IHG model, this assumption is made for each volume element, whose stability conditions are evaluated at different depths (typically per meter) until the bedrock or a stable layer is reached, thus defining the sliding surface characterized by the minimum safety factor value. The sliding surface can be shallow or even tens of meters deep [19,29], depending on where the conditions most unfavourable for stability are identified, cell by cell. The overall sliding surface can be irregular, since the volume elements may have different geometrical and physical–mechanical characteristics, variable in both depth and time. For instance, the soil water content and vegetation, affecting the slope response, are variable in space and time. The role of vegetation in the stability of shallow unsaturated soils is significant from the hydrological and mechanical point of view. It can be taken into account even by simplified approaches in limit-equilibrium analyses [30], as the IHG model does.

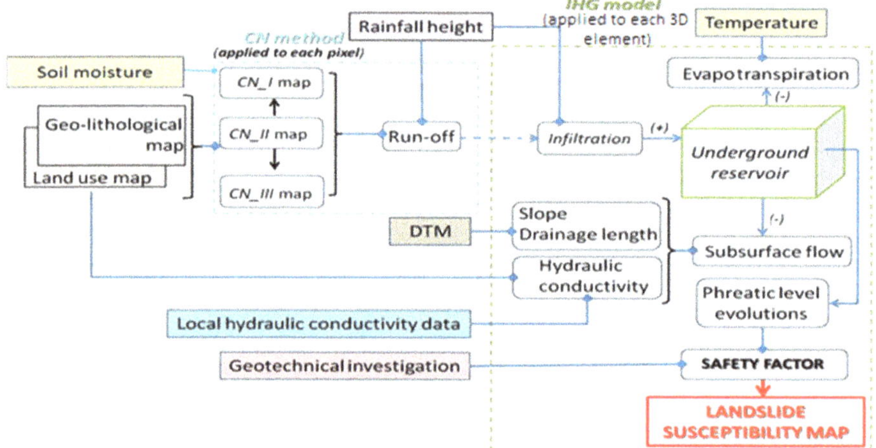

Figure 2. Flow chart of the IHG model.

The final products are maps of susceptibility to landslide failure (in raster format), both for measured and, if needed, forecasted rainfalls. The obtained maps are based on the spatial variation of the safety factor.

Within the AD-VITAM project, the IHG model is currently adopted for the analysis of five sites. Many analyses have already been carried out in relation to numerous past rainfall events on a daily scale, correlating the observations made on site by piezometers and rain gauges in order to calibrate the parameters of porosity and permeability to better capture the rise/fall phases of the water table as observed on site.

The low-cost, self-sufficient, remotely manageable monitoring network that feeds the IHG model comprises temperature and soil water content sensors; rainfall data are collected by rain gauges or meteorological radars, eventually integrated with other innovative techniques [31,32]. Moreover, Global Navigation Satellite System (GNSS) receivers are installed on poles fixed to the ground in order to check if the most critical areas, according to the IHG modelling, are actually subject to displacements. Obviously, direct validation is not possible, as the IHG model assesses the stability conditions through a global limit equilibrium method.

There are many reasons why soil moisture sensors are used in the LAMP system:

(1) The IHG model applies the Modified Curve Number Method (SCS 1972–1975) to calculate the amount of infiltrating rain. As mentioned above, the CN value depends on the land use and "antecedent soil moisture condition". Therefore, the measurement of the water content in the soil allows the hydrological balance to be determined.

(2) The IHG model evaluates the slope stability conditions by a three-dimensional limit-equilibrium analysis. The measurement of the soil water content profile over time is useful for the evaluation

of the stresses and the characterization of the mobilized soil strength in the shallow layers. In fact, partial saturation conditions influence the effective stresses, the mechanical behaviour of the soil [33,34] and, consequently, the stability conditions of a slope [35]. Therefore, measurements of soil water content are adopted in geotechnical stability analyses.

(3) The monitoring of the water content in the soil can also be useful for the analysis of soil moisture conditions immediately after emergencies. In fact, landslide risk conditions may persist long after the officially issued alert has ceased. By monitoring the soil water content (provided that the sensors have not been damaged during the emergency), it would be possible to understand when people, subjected to removal, can return to their homes safely.

Moreover, soil water content plays a crucial role in a wide variety of biophysical processes, such as seed germination, plant growth and nutrition. It affects water infiltration, redistribution, percolation, evaporation and greenery transpiration. Hence, for a proper description of the behaviour of the upper soil layers and, not least, for bio-engineering countermeasures to mitigate landslide risk, the assessment of soil moisture is also important.

2.2. Soil Water Content Sensors

Since much Italian territory is subject to soil landslides, the possibility of using easily replaceable/relocatable, autonomous, remotely controllable and low-cost sensors for monitoring widespread areas is particularly appealing. Obviously, the sensors have to be suitable for environmental monitoring (be little affected by temperature, salinity, soil texture, etc.) to guarantee satisfactory accuracy, precision and ease of use with their acquisition/release systems.

Focusing on the field measurement of soil water content, the monitoring network, recently installed in some of the sites defined in the abovementioned project, has been equipped with WaterScout capacitive probes (for details, see https://www.specmeters.com/weather-monitoring/sensors-and-accessories/sensors/soil-moisture-sensors/sm100), whose unit cost is about 75 Euros. They appear to be particularly suitable for monitoring soil slide areas and allow multiple installations (at different shallow installation depths, generally less than one meter) at each node of the settled networks. Figure 3 shows the main devices of the monitoring network, and Figure 4 shows the WaterScout SM100 used in relation to the FieldScout Soil Sensor Reader. Today, a complete soil moisture network consisting of five measurement nodes (with four WaterScouts per node) costs approximately 5000 euros in total, of which 30% relates to the sensors. These instruments are designed for agriculture, to optimize irrigation automation in plantations; however, with them having undergone considerable development, they have aroused interest in other fields of application. In civil engineering, for instance, they are suitable for hydrological–geotechnical analyses, as shown in this paper. It is worth pointing out that although the assembly of the monitoring system is relatively simple, its installation in a natural environment requires some care.

The described sensor provides both instantaneous and continuous volumetric water percentage readings, depending on whether the instrument is associated with the FieldScout Soil Sensor Reader rather than the Sensor Pup and, therefore, the monitoring network. The readings taken by the sensors and made available by the SensorPups and the FieldScout Soil Sensor Reader are based on the calibration curve provided by the manufacturer. Unfortunately, the type of soil on which this calibration was originally performed is not known because it is not declared by the manufacturer. Therefore, an appropriate experiment was carried out to evaluate the reliability of the WaterScouts and determine the soil-specific calibration laws.

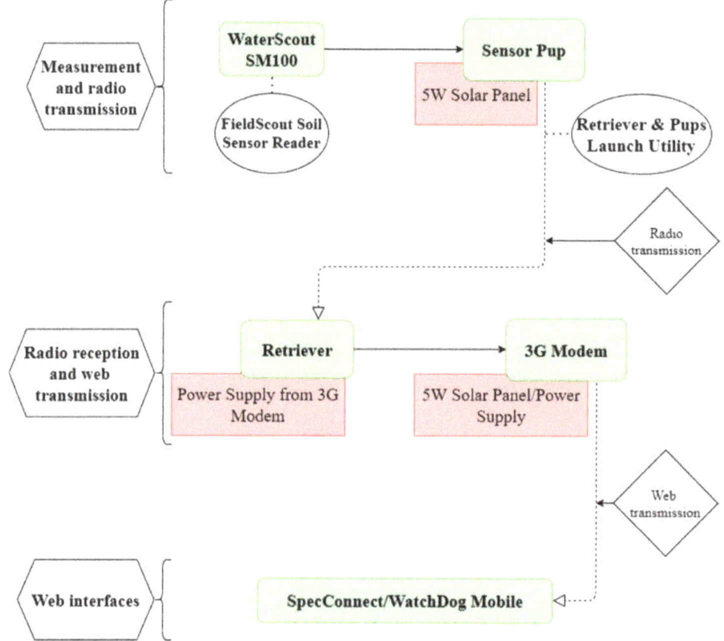

Figure 3. Flow chart of the monitoring network (Retriever & Pups Launch Utility: Version 1.2.7.3, Copyright 2016 Spectrum Technologies Inc., Aurora, IL, USA 60504).

(**a**)　　　　　　　　　　　　　　　　　　(**b**)

Figure 4. (**a**) The WaterScout SM100; (**b**) The FieldScout pocket Soil Sensor Reader.

2.3. Drill&Drop Probe and Sensor Distribution in the Test Field

Between 2018 and 2019, the University of Genoa carried out some research aimed at ascertaining the reliability of the WaterScouts SM100, based on a comparison between this sensor's response and the data provided by the capacitive probe Drill&Drop (Sentek Sensor Technologies), considering the latter as a reference tool.

The Drill&Drop probe (Figure 5) is a multi-capacitive probe (it has, in fact, one sensor per 10 cm of its length), which allows the measurement of the water content along a vertical alignment in the soil. It was previously studied in the laboratory [36], but being rather expensive, it was not used for monitoring the project sites.

Figure 5. The Drill&Drop probe.

Over a period of about three months, the two sensors (Drill&Drop and a set of WaterScouts) were installed close to each other, to monitor the evolution of the soil moisture profiles in a test field, up to a depth of about 90 cm below ground level.

The test field was monitored with eight WaterScouts, suitably identified by numbers and arranged along two verticals up to a depth of 85 cm, connected to two Sensor Pups (called PUP3 and PUP4), communicating with a Retriever. The WaterScouts used during the installation were selected by laboratory tests aimed at checking the validity of their responses. On that occasion (as well as in subsequent measurements in the test field), a type of signal drift occurred. This phenomenon was clearly appreciable in the laboratory when the sensors were immersed in water. In fact, each sensor was analysed in distilled water, i.e., in a medium with constant moisture, and, despite this, variation (sometimes even significant) in the sensor response was often observed. Subsequent analyses investigated the drift of the signal in water, attributable to the progressive formation of small air bubbles on the surfaces of the sensors.

The WaterScouts installed in the test field and the related Drill&Drop reference sensors, as well as their depths below ground level, are shown in Table 1.

The test field, for reasons of opportunity and accessibility, was set up in a site that, unfortunately, had rather peculiar soil properties. The first 30 cm consisted of soil, whose characteristics are indicated in Table 2, below which there was extremely heterogeneous landfill material containing stones, parts of bricks and small concrete debris. With reference to the latter portion of soil, it would have been necessary to define an appropriate calibration law for the Drill&Drop probe. Such calibration was not carried out, since the experiment with the Drill&Drop probe was only preliminary and preparatory for the experiment performed specifically for the WaterScout sensors installed on site.

Table 1. Sensors tested in the field.

WaterScout-PUP 3		WaterScout-PUP 4		Drill&Drop Sensors	
Number	Depth [cm]	Number	Depth [cm]	Number	Depth [cm]
54	−16.0	76	−8.0	2 (for PUP3), 1 (for PUP4)	−15.0
70	−31.5	69	−31.5	4	−35.0
21	−58.0	77	−58.0	6	−55.0
61	−85.0	78	−85.0	9	−85.0

Table 2. Characteristics of the test-field soil (the upper 30 cm). D_{50}: mean grain diameter (50% sieve passing); $C_u = \frac{D_{60}}{D_{10}}$: uniformity coefficient; G_s: specific gravity; n: porosity.

Test Field: Soil Characteristics (z = 0 cm ÷ 30 cm)	
D_{50} [mm]	1
C_u [−]	16
Average organic content [%]	7.8
G_s [−]	2.6
n [%] (z = −18 cm ÷ −28 cm)	43

2.4. Calibration Procedure

As specified in [1], one of the main limitations of the capacitive sensors is the operating frequency, which appears to be around 70–80 MHz. This makes the response of the moisture sensors very sensitive to the presence of clay in soils, for which a clear deviation of the calibration curve from that in standard soil conditions is expected. Therefore, instrument-specific calibration is required to overcome this limitation.

The soil-specific calibration was carried out on samples taken very close to the verticals passing through the nodes of the monitoring network. The calibration is still in progress or, even, to be carried out for some of the five monitored sites.

In Section 3.5, attention is focused on the monitoring network installed in Ceriana-Mainardo, next to whose nodes soil samples were taken (in analogy with what was done for the other sites) for the pertinent sensor calibration discussed here.

Since a large number of soil samples was required for the calibration of the sensors (about ten for each site), it was decided to reduce the costs by using equipment much more economical than a classical one. Samplers made from PVC-UH pipes (diameter, 110 mm; thickness, 0.4 mm) were equipped with special elements made of steel: a distributing crown to be mounted on top of the sampler and an inner-flush lower ring with a cutting edge, to facilitate the sampler's penetration into the soil with hammering (Figure 6) and to avoid the breakup of local samplers.

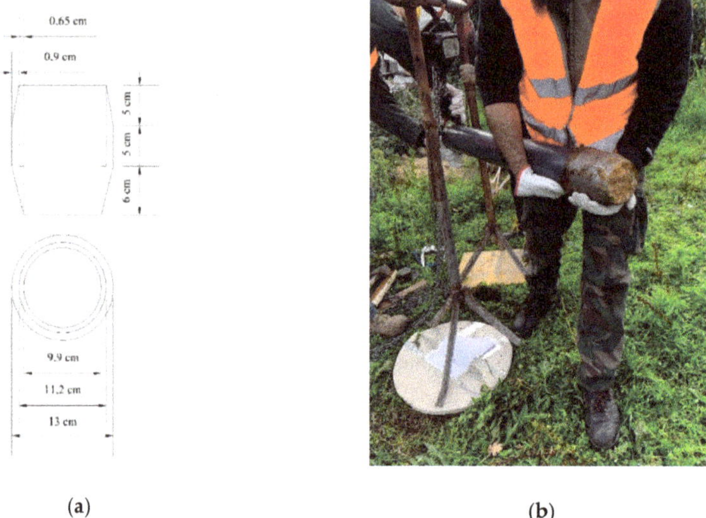

(a)   (b)

Figure 6. (a) Details of the specifically developed lower ring with a cutting edge; (b) The components as mounted on the sampler during on-site collection.

At the site of Ceriana (Liguria, Italy), more precisely, in the Mainardo locality (very close to the main village), sampling had already been carried out near to all of the five nodes of the installed network.

In the laboratory, with the extracted soil samples, we proceeded with:

(1) The geotechnical characterization of the soil: the grain size, porosity, particle density and gravimetric water content;

(2) The determination of soil-specific calibration curves; soil moisture sensors provide data in terms of the volumetric water content θ_V [%], which is related to the gravimetric water content w [%] through Equation (1)

$$\theta_V = w \cdot \frac{\rho_s}{\rho_W} \qquad (1)$$

where ρ_s is the dry soil density (given by the ratio of dry soil mass to the relevant volume); ρ_W is the density of water; w is the gravimetric water content (given by the ratio between the mass of water in the soil to the mass of dry soil).

The procedure adopted for the calibration, with reference to each given sample moisture content, varied to obtain a calibration curve, can be summarized as follows:

(1) A portion of soil is placed in a container (Figure 7a) with known mass (Tare) and internal volume (V);

(2) The soil and container are weighed (Figure 7b) so that the wet sample density ρ_{wet} can be calculated by Equation (2), where m_i is the difference between the mass of the wet soil and container assembly, and the Tare.

$$\rho_{wet} = \frac{m_i}{V} \qquad (2)$$

(a) (b)

Figure 7. (a) Container; (b) Weighing of the wet soil sample.

(3) Four WaterScouts are inserted into the sample (Figure 8), and the raw data of the soil water content are collected from each sensor by the FieldScout Soil Sensor Reader.

(a) (b)

Figure 8. (a) Insertion of the first moisture sensor into the wet sample; (b) Final arrangement of the four sensors.

(4) A portion of the sample is then taken for the measurement of the gravimetric water content w, which allows the dry mass of the sample m_s to be determined with Equation (3); the apparent soil density ρ_s is determined by Equation (4).

$$m_s = \frac{m_i}{w+1} \tag{3}$$

$$\rho_s = \frac{m_s}{V} \tag{4}$$

This procedure makes it possible to calculate, by means of Equation (1), the volumetric water content of a specific sample and to record the raw data associated with it, as the averages of the four output data of the sensors used in Step 3. The repetition of the above steps on soil samples with different moistures produces a series of data useful for determining the correlation between the raw data and soil moisture content.

It is appropriate, however, to highlight some fundamental aspects related to the nature of the raw data or the output voltage V_{out} of the sensors. First, it varies depending on the supply voltage V_{in} of the sensors, which ranges from 3 to 5 V, and the device one decides to connect to the instrument. In fact, while the Sensor Pup supplies the sensors with an input voltage of about 3 V, the FieldScout Soil Sensor Reader supplies the sensors with that of about 5 V; consequently, the output voltages measured by the two devices are different.

Nevertheless, through an analysis of the A/D converters of the devices, it was possible to define the relationship between the two raw data.

$$RAW\ Sensor\ Pup\ [mV] = 0.73 \cdot RAW\ FieldScout\ [mV] \tag{5}$$

Please note that the actual raw data used in calibration are not the V_{out} readings but the V_{out}/V_{in} ratios. The statistical characteristics of Equation (5) will be discussed in Section 3.2.

2.5. Field Installation

The field installation of the monitoring network starts with an inspection aimed at choosing the points at which to place the sensors and the devices necessary for transmission and reception. The Retriever must be able to pick up the radio signals of the Sensor Pups distributed in the area, and the Modem linked to it must connect to a GSM operator that ensures reliable network coverage. Therefore, the choice of the node locations necessarily depends both on the presence of a small area in which the sensors can be inserted and on the radio signal cover; the latter can be evaluated by connecting the Retriever to a PC via a USB cable and using the appropriate Retriever/Pups interface.

With reference to each measuring node, the sensors are installed in the ground at different depths in order to describe the vertical moisture profile in the upper 90 cm.

There are two main modes of installation. The fastest one involves inserting each sensor vertically at the desired depth through a pre-installation hole (Figure 9). If this operation is hampered by the presence of stones/pebbles or by meeting a particularly compact soil level, it is necessary to opt for horizontal installation from a trench front (Figure 10).

For the insertion of the WaterScouts in the ground, a PVC pipe is used; the sensor is pushed into the soil along the full probe length, being inserted at the pipe end opening, while its wire exits from the other side; the pipe is then removed when the sensor is properly set.

Whichever installation method is chosen, the sensor surfaces must be in full contact with the soil. Sometimes (especially if the soil to be penetrated by the sensor is particularly stiff), the cuttings, coming out during the drilling phase, are applied directly on the surface of the instrument, which is then simply inserted at the bottom of the pre-drilled hole without pushing, in order to avoid the rupture of the sensor. The functionality of any installed sensor must always be immediately checked, by using the FieldScout Soil Sensor Reader.

The vertical installation involves, after the positioning of the sensors, the application of a certain quantity of dry bentonite in the upper part of the drilled hole, then moistened by pouring water from the top, to seal the installation hole to avoid preferential infiltration paths for the surface runoff water.

(a)   (b)   (c)

Figure 9. Vertical installation: (**a**) Creation of the vertical hole; (**b**) Positioning of the WaterScout in the PVC pipe; (**c**) Insertion of the sensor in the hole.

Figure 10. Example of a horizontal installation.

Horizontal installation requires the insertion of the WaterScouts in horizontal pre-holes a few centimetres deep. In this case, it is not necessary to use bentonite. Despite this, the opening of a temporary excavation is needed, and once installation is complete, the subsequent backfilling alters the initial conditions with reference to water circulation. Therefore, before backfilling, it is necessary to apply a waterproof sheet on the excavation front where the sensors have been installed.

As mentioned, the Sensor Pups distributed in the area are powered by 5 V solar panels, while the Modem is connected to the power net with a transformer, although it is also possible to power it with a solar panel. The Retriever is powered by the Modem itself, through the same AUX cable used for the data transmission.

The monitoring devices are installed in a natural environment. Therefore, it is advisable to use protective sheaths coated with corrugated aluminium to prevent wild rodents from damaging the sensor cables. It is also advisable to complete the installation of the Sensor Pups and attached sensors with the construction of a small protective fence that locally surrounds the control unit support pole.

The described measuring instruments do not conflict with other monitoring devices in field, such as the GNSS used to acquire the surface displacements.

It is worth noting that the procedures relating to the calibration and field installation of the water content sensors defined by us take into account what is indicated in the Sentek Tec. and Spectrum Tec. manuals [37–40].

2.6. Case Studies

In July 2019, the inspections to decide the monitoring network nodes on the AD-VITAM project sites began (Figure 11a). Between November and December 2019, the monitoring networks at Mendatica and Ceriana-Mainardo (Figure 11b) (Italy) were completed, while the installation in Vence (France) was concluded in February 2020.

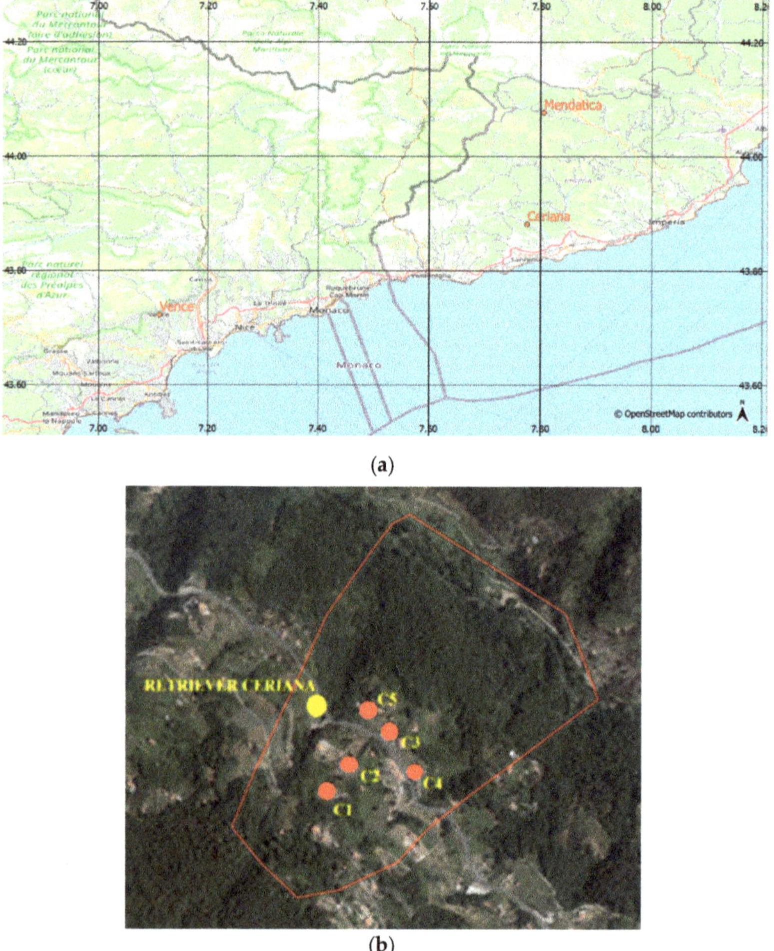

(**a**)

(**b**)

Figure 11. (**a**) Location of Ceriana, Mendatica and Vence (map in WGS84); (**b**) Installed monitoring network. The nodes (red dots) are denoted by C1–C5.

To establish the positions of the nodes, all the Sensor Pups (five for each landslide site) were activated so as to enhance the radio signal, determined by the "bridge" effect between the different control units distributed over the area. The instruments were provisionally fixed on telescopic support poles, which were placed in the planned points for ascertaining the connection between the nodes.

Furthermore, it was necessary to find a suitable point for the positioning of the Retriever. Although the installation points can be chosen in advance through analysing the cartography of the territory, it is not wise to forgo a field evaluation. In fact, the nodes must be positioned so as to allow good data transmission. Once the location of the nodes was established, how to install the WaterScout sensors according to the soil properties was decided.

The deposits found in Mendatica showed the presence of boulders and particularly dense soil layers, which have repeatedly hindered the insertion of the sensors at the provided depths. Hence, at two of the five installation points, it was necessary to carry out horizontal insertion. On the contrary, the Ceriana site presented favourable installation conditions, with the exception of a single point, where the soil layers below 70 cm became very stiff, such that one of the sensors broke. The realization of the pre-holes for the installation of the WaterScouts is also useful for determining if sampling near the node will be simple. Finally, the Vence site in France (Figure 11a) presented favourable installation conditions. However, even though the deposits did not show a marked presence of boulders, the soil density was such as to contrast the sensor insertion. For this reason, it was decided to apply the soil cuttings directly on the sensor surfaces, as described above; the sensors were then vertically placed at the desired depth.

In the following, there will be exclusive reference to the results from the measurement node named C5 located in Ceriana, as the pertinent calibration of the WaterScouts had already been carried out. The soil water content measurements at node C5 (Figure 11b) will be linked to four rainfall events that occurred between May and June 2020.

3. Results

The following results are related to the behaviour of moisture sensors with the rainfall-induced changes in soil water content in the test-field scenario, described in Section 2.3. The results useful for the definition of Equation (5) are also described. Finally, some evaluations are reported of the characterization of the soil taken from one of the measurement nodes of the Ceriana-Mainardo site, as well as the results of the calibration described in Section 2.4 applied to it. Again, referring to the abovementioned node C5, the first water content measurements obtained through the described monitoring systems are then presented.

3.1. Results of the Preliminary Study on the Test-Field: Comparison between WaterScout SM100 and Drill&Drop Probe

The graphs in Figures 12 and 13 compare the soil water content measurements made by the sensors during the rainfalls that occurred between 29 December 2018 and 19 March 2019 at the test field. It is worth recalling that the test field was characterized by a very shallow soil layer up to a depth of 30 cm (sparsely vegetated with grass) and extremely heterogeneous and heterometric landfill material (containing stones, parts of bricks and small concrete debris) below.

At first glance, the measurements obtained through the WaterScouts appeared to be significantly different from the moisture detected by the Drill&Drop probe, both in terms of the measured values and in terms of their variations. Drift of the signal was particularly evident at depths between 55 and 85 cm, where the soil was not significantly affected by water infiltration and evapotranspiration phenomena. Figure 14 (in accordance to [41]) shows the presence of a zone of almost constant moisture in relatively deep layers of soil and a relatively shallow zone, where a cyclic variation of moisture contents occurs. The profiles shown in Figure 14 are merely qualitative. In fact, the soil moisture profiles are influenced by the specific conditions of the site; they depend on many factors, including the characteristics and texture of the soil, the vegetation and the climate.

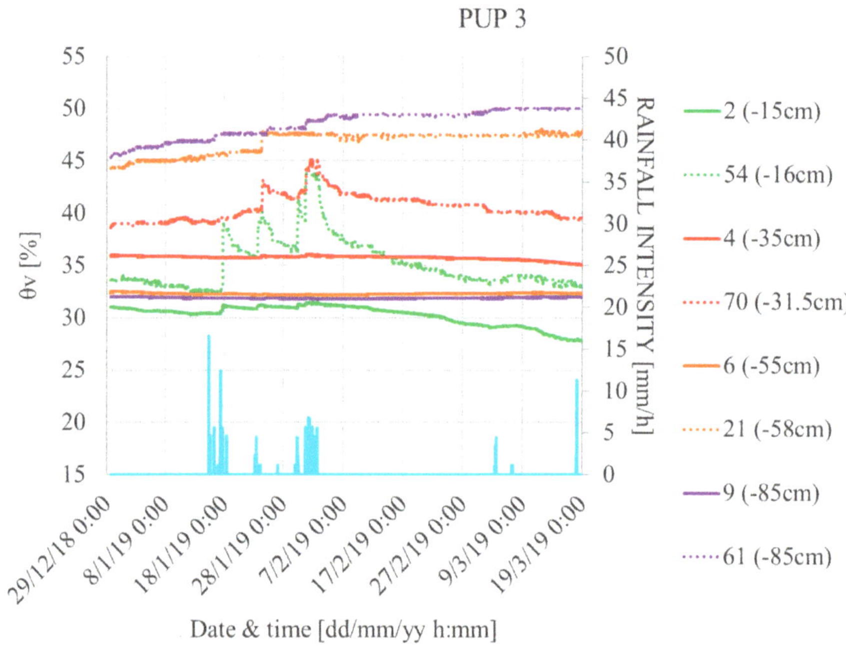

Figure 12. Measurements by WaterScout n.54, n.70, n.21 and n.61 and by the pertinent reference sensors n.2, n.4, n.6 and n.9 of Drill&Drop for 29 December 2018 to 19 March 2019.

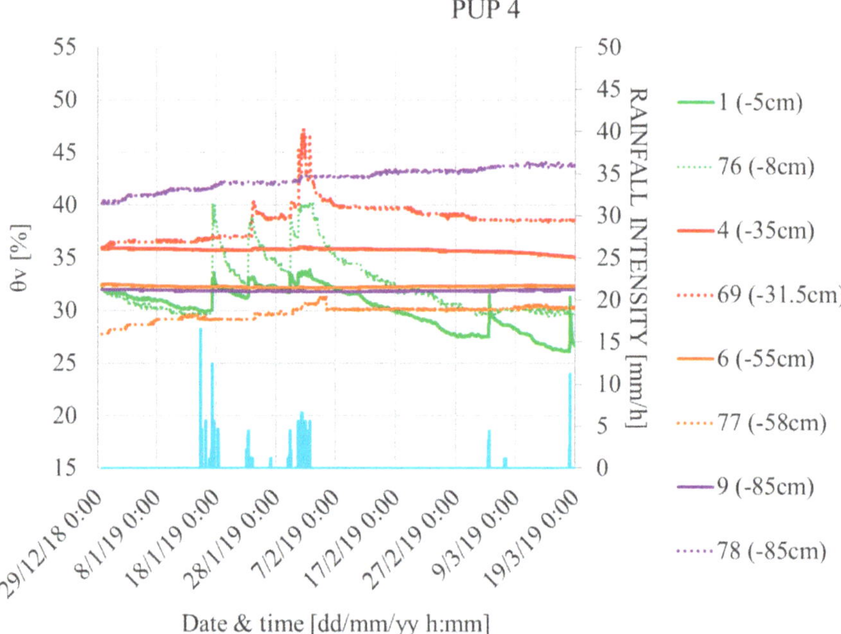

Figure 13. Measurements by WaterScout n.76, n.69, n.77 and n.78 and by the pertinent reference sensors n.1, n.4, n.6 and n.9 of Drill&Drop for 29 December 2018 to 19 March 2019.

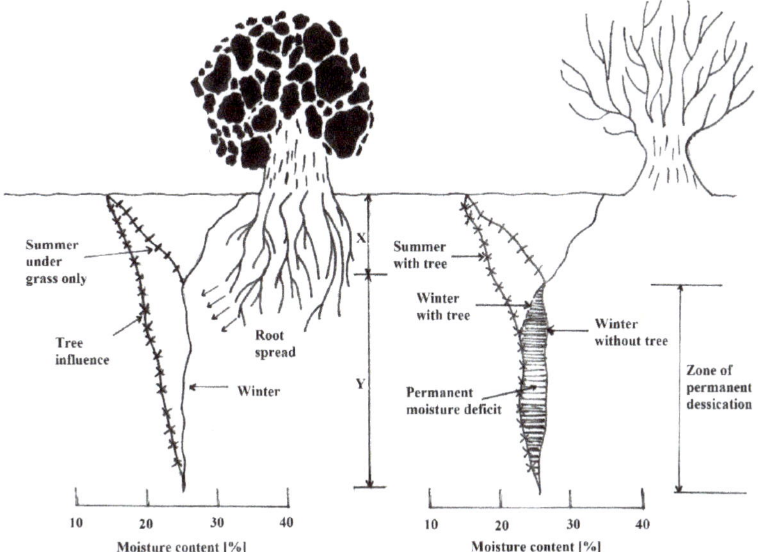

Figure 14. Changes in soil moisture content: tree influence (summer) on the left and tree influence (winter) on the right.

Figures 12 and 13 show a substantial difference between the data measured by the two probes under examination. The response of the Drill&Drop sensors is quite flat, and this would have deserved further investigation. On the other hand, the measurements of the WaterScouts were not calibrated by a soil-specific law, suggesting the need to focus subsequent research on the laboratory calibration of the WaterScouts.

The measurements in the test field also indicate that close attention should be paid to the installation of sensors. Therefore, even greater care was dedicated to the subsequent installation of the sensors on site. In fact, correct installation and good soil–sensor contact are extremely important.

Since the WaterScouts are used in the monitoring networks, attention was focused on these probes, as shown in the following.

3.2. Results of the Study about the Relationship between the FieldScout Soil Sensor Reader and Sensor Pup Raw Data

Equation (5) is the result of a specific analysis of the FieldScout Soil Sensor Reader used during the calibration and installation phases. This analysis shows that the power voltage supplied by the instrument is a little more than 4 V and slightly lower than the maximum supply voltage of 5 V specified in the user manuals. Considering, therefore, the supply voltage of the Sensor Pup being equal to 3 V and 12-bit A/D conversion, Table 3 and Figure 15 show the relationship between the two raw data expressed by Equation (5), which perfectly fits the data with a coefficient of correlation $R^2 = 1$. This approach led to a very precise relationship between the Sensor Pup and FieldScout raw data.

Table 3. Results of the A/D conversion.

RAW FieldScout [mV]	RAW Sensor Pup [mV]
1	0.73
500	366.21
1000	732.42
1500	1098.63

Table 3. *Cont.*

RAW FieldScout [mV]	RAW Sensor Pup [mV]
2000	1464.84
2500	1831.05
3000	2197.27
3500	2563.48
4096	3000.00

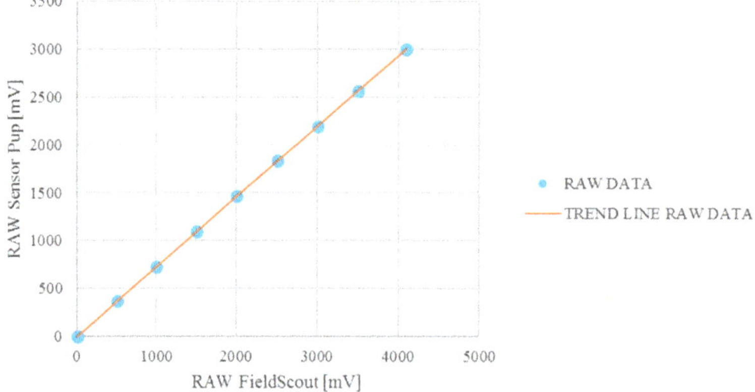

Figure 15. Relationship between FieldScout Soil Sensor Reader and Sensor Pup raw data.

3.3. Results of the Characterization of the Soil Sampled in Ceriana-Mainardo for the Calibration Procedure

The described calibration operations were performed on the soil pertinent to a borehole realized at the node, called C5, of the monitoring network in Ceriana-Mainardo (Imperia, Italy). The landslide deposit is composed of heterogeneous and heterometric material.

The core samples, retrieved from the upper 60 cm of the soil, did not show substantial variability with depth. The calibration of the soil moisture sensors confirmed this observation. Therefore, it was decided to report the properties of a single portion of the extracted soil, representative of the material between 35.5 cm and 45.5 cm depth below ground level (Figure 16). This portion was extracted from the core sample, limiting the disturbance in order to evaluate the porosity n, the voids index e and, therefore, the relative density Dr (Table 4).

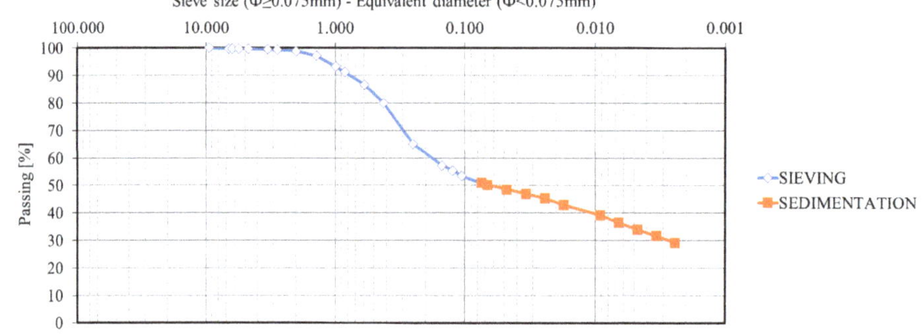

Figure 16. Grain size curve.

Table 4. Characteristics of the soil at C5. D_{50}: mean diameter; $C_u = \frac{D_{60}}{D_{10}}$: uniformity coefficient; Gs: specific gravity; e: void index; e_{max}: maximum void index; e_{min}: minimum void index; Dr: relative density.

Ceriana-Mainardo C5: Characterization of the Soil (z = 0 cm ÷ 30 cm)	
D_{50} [mm]	0.0679
Cu [−]	60
Gs [−]	2.6
e [%]	82.9
e_{max} [%]	128.7
e_{min} [%]	59.4
Dr [%]	66

3.4. Calibration Results for the Soil Sample Taken from Ceriana-Mainardo

The first results of the calibration procedure for the soil at the node C5 in Ceriana-Mainardo (IM) are shown in Figure 17. By interpolating the data by linear regression, it is possible to obtain the calibration correlation shown in Table 5. Figure 17 also shows the comparison between the calibration function exposed in Table 5 and the relationship provided by Spectrum Tec., which is not linear. The validity range for the calibration function (named UNIGE CALIBRATION in Figure 17) expressed in Table 6 is $0.32 \le \frac{V_{OUT}}{V_{IN}} < 0.47$.

Although the experimental data (Figure 17) suggest that two trends can be identified, a single calibration equation is introduced in the interest of simplicity, which represents the overall trend of the data. In fact, upon analysing other possible calibration laws (e.g., a segmented linear regression and second-degree polynomial curve), the results did not show any significant improvement compared to the simple trend line indicated in Table 5.

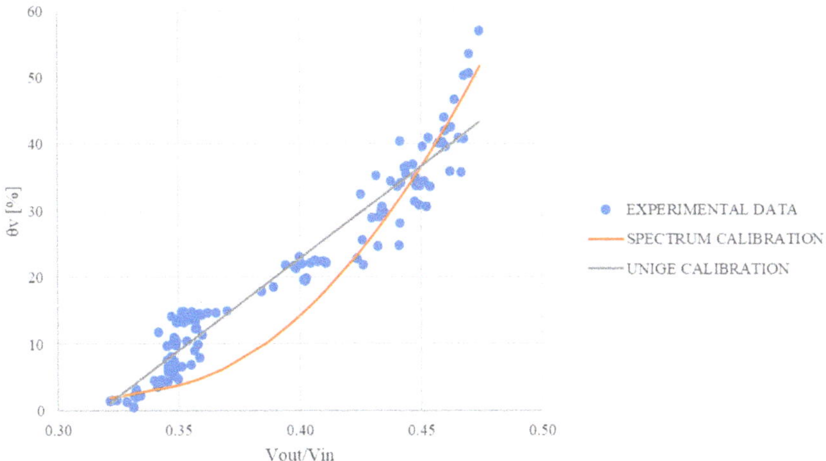

Figure 17. Comparison between the Spectrum Tec. calibration law (SPECTRUM CALIBRATION) and the soil-specific calibration law determined by the University of Genoa (UNIGE CALIBRATION).

Table 5. Calibration function. R^2: coefficient of determination; RMSE: root mean square error.

Equation	Number of Data	a	b	R^2	RMSE [%]
$\theta_V[\%] = a\,\frac{V_{OUT}}{V_{IN}} + b$	122	276.5	−87.7	0.92	3.7

3.5. Rain Data and Water Content Measurements in Ceriana-Mainardo

In November 2000, Ceriana-Mainardo was subject to shallow instabilities, which particularly involved the Provincial Road n. 44 and the adjacent areas (Figure 18a). Geognostic investigations (continuous core drillings) and geotechnical and geophysical surveys were carried out. Laboratory soil characterizations, light dynamic penetrations (type Penny30), inclinometers and geophysical tests (seismic refraction alignments and down-holes), allowed the characterization of the subsoil sequences. Monitoring the site with piezometers and rain gauges made it possible to assess the water table fluctuations and occurring rainfalls, respectively.

The soil moisture monitoring network installed at the Ceriana-Mainardo site has been in operation since December 2019. At the WSN nodes, distributed as shown in Figure 18b, soil water content values are recorded every 5 min. Four measuring points were installed at each node. At node C5 (represented in Figure 11b), the sensors are placed at 10 cm, 35 cm, 55 cm and 85 cm below ground level. The sensors are similarly located in the other four nodes. All the sensors in Ceriana-Mainardo were installed vertically. Node C5 is shown in Figure 19a,b.

In Ceriana-Mainardo, two core samples were taken (similarly to how samples were taken in Mendatica and Vence). They were adjacent to each of the measuring nodes; only one sampler was broken on site. Figure 19c shows the sampling at node C5.

The measurements of the water content in the soil at node C5 during some rainfall events (Tables 6–8) are presented below; at C5, the steady water table fluctuates around 21 m below ground level. Figures 20a, 21a and 22a show the trends in water content measured at various depths. The represented soil moisture values were calibrated by the regression law indicated in Table 5 denoted by the acronym UNIGE. The graphs allow the understanding of how the water content varies over time at the four depths where the sensors are positioned. The rainfalls that occurred are also represented.

Figure 20b,c, Figures 21b and 22b show the variations of the vertical moisture profiles obtained by the UNIGE calibration law. Three profiles are represented; they are referred to three time-windows: before the beginning of the rainfall (undisturbed conditions), immediately after the rainfall peak, and at the end of the event.

These first measurements (Figures 20–22) appear encouraging and rather consistent with the hydrological processes on site. However, further measurements at C5 and the other nodes will allow a proper understanding of the water-content changes in the soil.

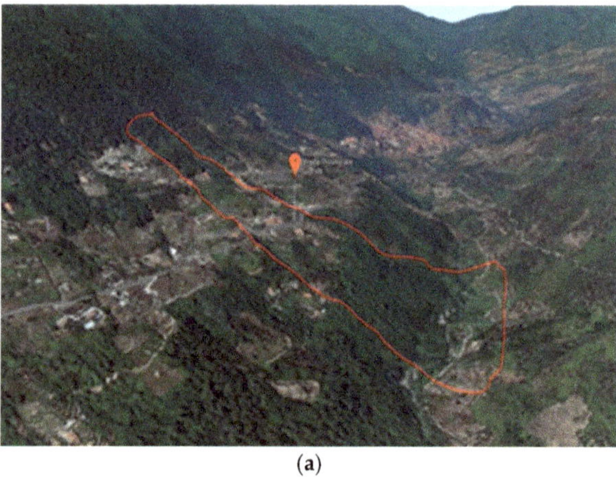

(a)

Figure 18. *Cont.*

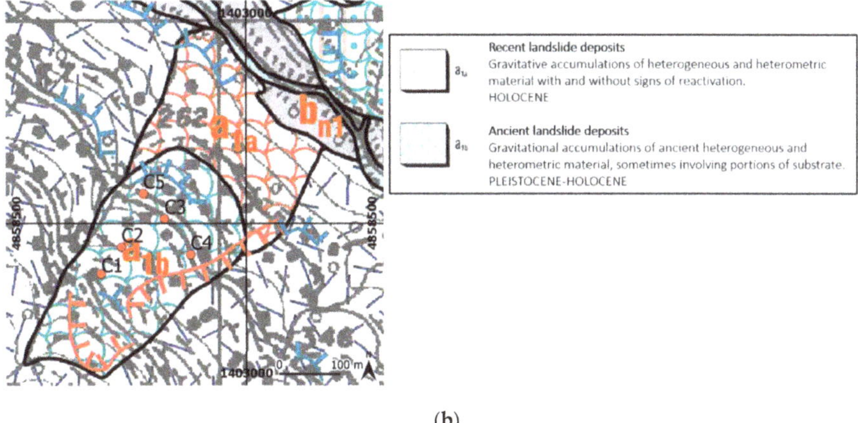

(**b**)

Figure 18. Ceriana-Mainardo: (**a**) Landslide perimeter view overlapped in 3D; (**b**) Geological map. The landslide perimeter (red polyline) and the nodes C1–C5 (red dots) are reported.

(**a**)　　　　　　　　　　　(**b**)　　　　　　　　　　　(**c**)

Figure 19. C5 node: (**a**) Node identification and warning signboard; (**b**) Carrying and rising pole of both the Sensor Pup and the solar panel; (**c**) Soil sampling.

Table 6. Characteristics of the 16–17 May 2020 rainfall events.

CERIANA-370 m		
Characteristics	**16 May 2020 EVENT**	**17 May 2020 EVENT**
MEAN INTENSITY [mm/h]	1.3	2.3
MAXIMUM INTENSITY [mm/h]	2.4	4.2
CUMULATE [mm]	8.8	11.4
DURATION [h]	7	5

Table 7. Characteristics of the 3–4 June 2020 rainfall events.

CERIANA-370 m, 3–4 June 2020 EVENT	
MEAN INTENSITY [mm/h]	4.5
MAXIMUM INTENSITY [mm/h]	9.6
CUMULATE [mm]	80.0
DURATION [h]	18

Table 8. Characteristics of the 13 June 2020 rainfall event.

CERIANA-370 m, 13 June 2020 EVENT	
MEAN INTENSITY [mm/h]	8.4
MAXIMUM INTENSITY [mm/h]	27.8
CUMULATE [mm]	67.4
DURATION [h]	8

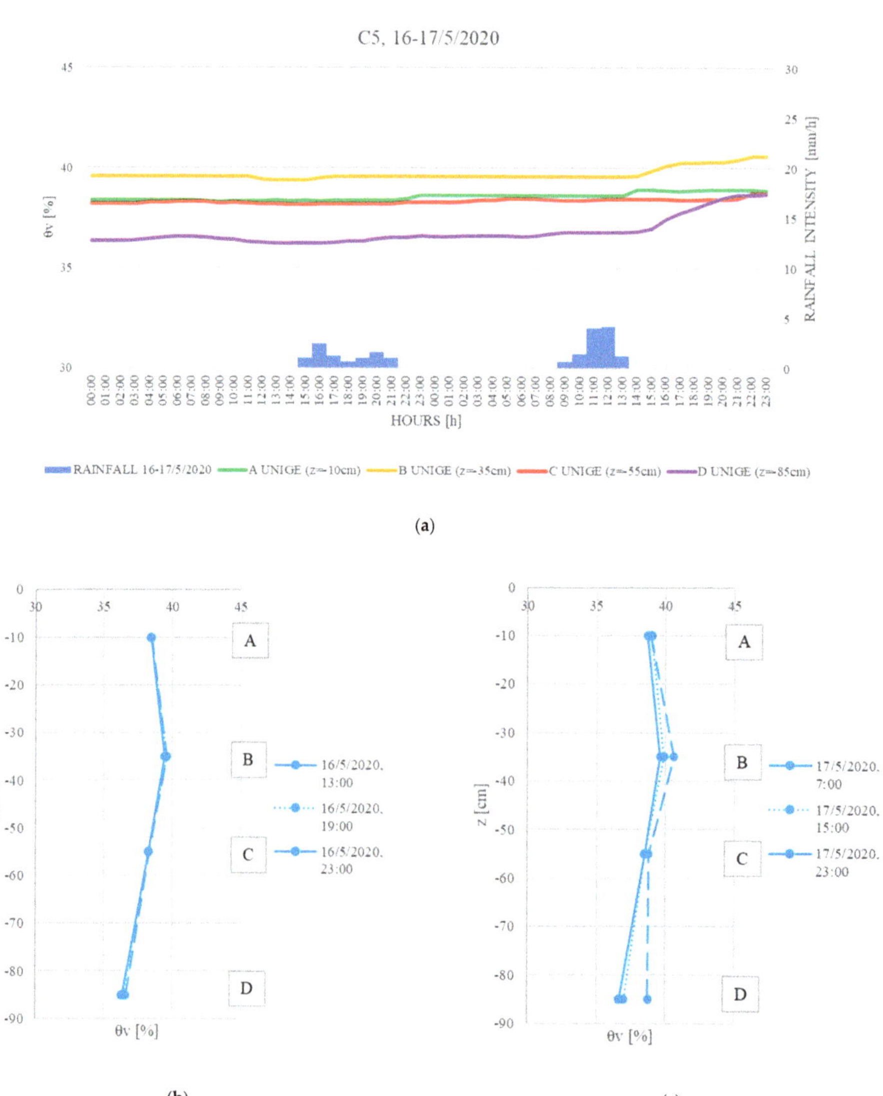

Figure 20. (**a**) May 2020 rainfall events and soil water contents; (**b**) Vertical soil moisture profiles 16 May 2020; (**c**) Vertical soil moisture profiles 17 May 2020.

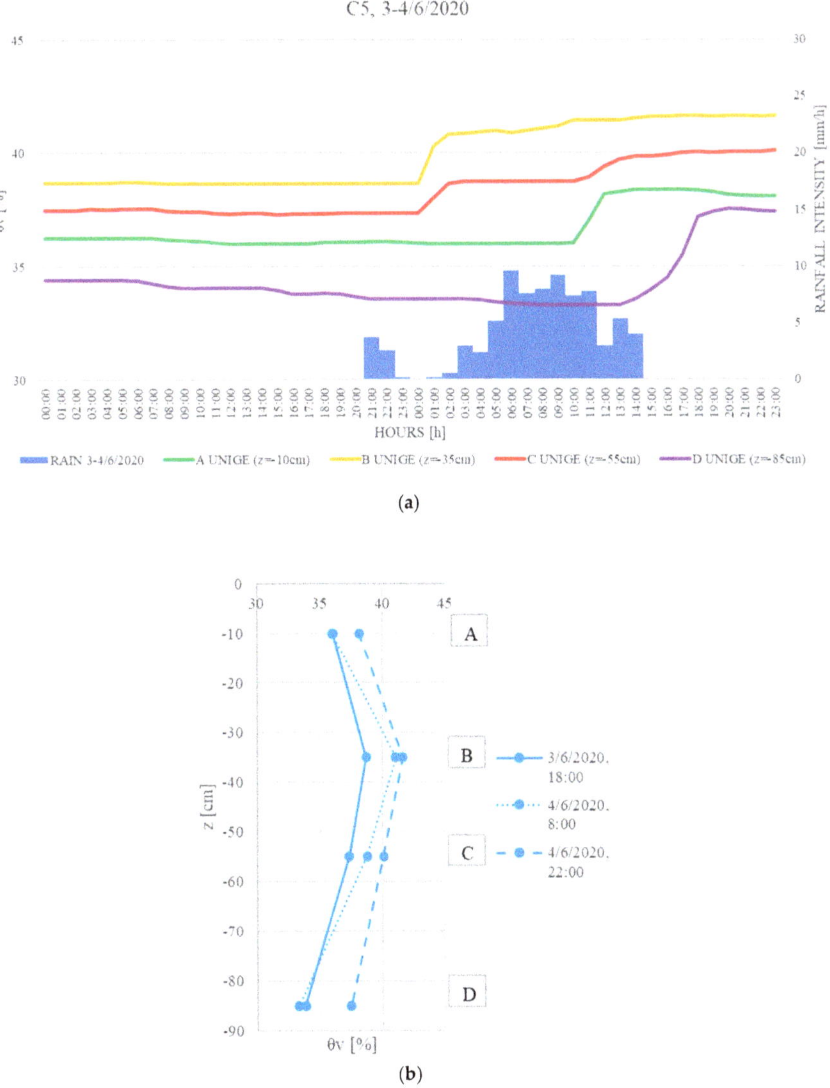

Figure 21. (**a**) 3rd and 4 June 2020 rainfall events and soil water contents; (**b**) Vertical soil moisture profiles for the 3–4 June 2020 rainfall events.

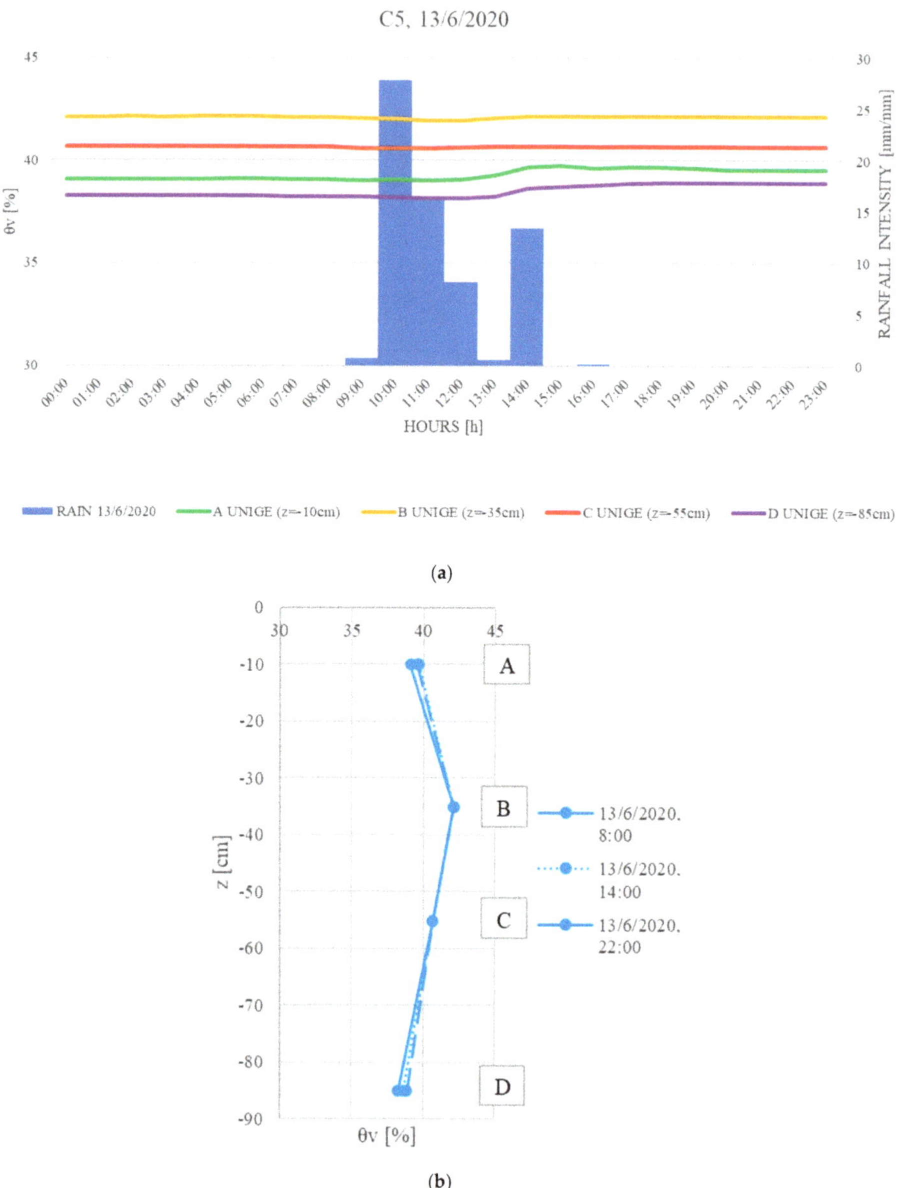

Figure 22. (**a**) 13 June 2020 rainfall event and soil water contents; (**b**) Vertical soil moisture profiles Figure 13. of June 2020 rainfall event.

4. Discussion

The water content measurements reported in this paper were taken at a single node of one of the five sites studied in the AD-VITAM project (the C5 at Ceriana-Mainardo). These measurements correspond to a short period, since the monitoring network was installed a few months ago.

The measurements were suitably calibrated because of the experiments performed on the soil taken on site at this node.

The data obtained by the field measurements are encouraging. In fact, Figures 20a, 21a and 22a not only show good behaviour for the sensors, which continuously detect variations in soil moisture, but also indicate that no preferential infiltration paths were generated during the installation phases. In fact, as shown in those figures, the increases in water content do not occur immediately after the rainfall but are recorded a few hours after the rain peaks. Figure 20b,c, Figures 21b and 22b clearly show the changes in water content at shallow depths when rainfall occurs. Thus, the measurements over time and soil water content profiles seem to capture the soil moisture dynamics, demonstrating their usefulness for describing the slope response to rainfalls.

Full analysis will be possible when water content measurements become available at all nodes and when the database becomes sufficiently rich and representative. All five sites will be equipped with the LAMP system by the end of 2020. This will allow the study of the response of the slopes to the rainfalls, both in real time (thanks to the WSN) and for prediction (if the short-term forecasted rains are considered), as well as improving landslide risk management, the design of mitigation measures and territorial planning [42].

5. Conclusions

The LAMP system was formulated for the analysis and forecasting of susceptibility to landslides triggered by rainfalls. It is based on an integrated hydrological–geotechnical model that is physically based and fed by a low-cost, self-sufficient, remotely manageable and easily replaceable/relocatable monitoring network. WaterScout capacitive sensors are positioned at sites potentially susceptible to landslides, at shallow depths (typically less than 1 m).

Knowledge of the near-surface water content over time is needed for hydrological and geotechnical analysis, the study of the hydro-mechanical contribution from the vegetation and, eventually, landslide risk mitigation with soil bio-engineering countermeasures. Additionally, the monitoring of the soil water content may be useful for analysing the soil moisture conditions immediately after emergencies, to determine when the risk conditions no longer exist.

This paper describes the installation phases for the monitoring networks, pinpointing the care needed in the choice of nodes, the arrangement of the sensors and, not least, the protection of the materials left on site. Particular attention has been paid to experiments conducted to evaluate the reliability of the used sensors and calibrate them. Soil-specific calibration was carried out in the laboratory on soil samples taken very near to the WSN instrumented verticals.

WaterScout calibration is very important. The WaterScout-SM100's operating frequency range is 70–80 MHz, and relevant variations in the sensor response may occur, especially in clayey soils. A soil-specific calibration is particularly recommended, to take into account the characteristics and textures of the in situ soil.

Although most soil moisture sensors are sensitive to temperature and salinity, the WaterScouts are little affected by these factors [3], so they can be adopted for effective environmental monitoring.

Since these probes must be inserted into the soil and be in contact with it, their use is possible in soils that can be penetrated, and the immediate confirmation of full soil–sensor contact with a FieldScout Soil Sensor Reader is recommended. In fact, such a reading allows the operator to immediately determine if the sensor has made full contact with the soil; a measurement close to zero would indicate partial contact with the air.

These sensors, providing continuous and real-time data, are substantially low in cost, easily relocatable and replaceable. Moreover, a complete and remotely manageable soil moisture monitoring network is relatively inexpensive, does not require onerous preliminary operations and is not bulky or unsightly. A monitoring network made up of five measurement nodes can be installed by three expert operators in only 2–3 days.

The soil water content measurements, so far acquired by the recently installed monitoring networks, are very satisfying and encouraging. The integration of such monitoring system in the LAMP system may be useful for the analysis and prediction of landslides triggered by rainfalls and could be of real support in risk management.

Author Contributions: Conceptualization, R.B., B.F. and R.P.; methodology, R.B., B.F., A.I. and R.P.; software, R.B., B.F., A.I. and R.P.; validation, A.I.; formal analysis. A.I.; investigation, R.B., B.F., A.I. and R.P.; resources, R.B., B.F., A.I. and R.P.; data curation, B.F. and A.I.; writing—original draft preparation, R.B., B.F., A.I. and R.P.; writing—review and editing, R.B., B.F., A.I. and R.P.; visualization, R.B. and A.I.; supervision, R.B. and R.P.; project administration, R.B. and B.F.; funding acquisition, R.B., B.F. and R.P. All authors have read and agreed to the published version of the manuscript.

Funding: This research was funded by Interreg V-A France-Italie, ALCOTRA "AD-VITAM" 2014-2020, Axis 2, Safe Environment programme, grant number 1573. The APC was funded by Roberto Passalacqua's research funds from the University of Genoa.

Acknowledgments: The authors acknowledge the support given by AD-VITAM Partners. A particular acknowledgement goes to the Unione dei Comuni della Valle Argentina e Armea and to the lab technicians of the Department of Civil, Chemical and Environmental Engineering of the University of Genoa.

Conflicts of Interest: The authors declare no conflict of interest. The funders had no role in the design of the study; in the collection, analyses or interpretation of data; in the writing of the manuscript; or in the decision to publish the results.

References

1. Tarantino, A.; Pozzato, A. Strumenti per il monitoraggio della zona non satura. *Riv. Ital. Geotec.* **2008**, *3*, 109–125.

2. Nagahage, E.A.A.D.; Nagahage, I.S.P.; Fujino, T. Calibration and Validation of a Low-Cost Capacitive Moisture Sensor to Integrate the Automated Soil Moisture Monitoring System. *Agriculture* **2019**, *9*, 141. [CrossRef]

3. Adla, S.; Rai, N.K.; Sri Karumanchi, H.; Tripathi, S.; Disse, M.; Pande, S. Laboratory Calibration and Performance Evaluation of Low-Cost Capacitive and Very Low-Cost Resistive Soil Moisture Sensors. *Sensors* **2020**, *20*, 363. [CrossRef] [PubMed]

4. Iverson, R.M. Landslide triggering by rain infiltration. *Water Resour. Res.* **2000**, *36*, 1897–1910. [CrossRef]

5. Crosta, G.B.; Frattini, P. Distributed modelling of shallow landslides triggered by intense rainfall. *Nat. Hazards Earth Syst. Sci.* **2003**, *3*, 81–93. [CrossRef]

6. Vassallo, R.; Mancuso, C.; Vinale, F. Effects of net stress and suction history on the small strain stiffness of a compacted clayey silt. *Can. Geotech. J.* **2007**, *44*, 447–462. [CrossRef]

7. Mancuso, C.; Vassallo, R.; d'Onofrio, A. Small strain behavior of a silty sand in controlled-suction resonant column - Torsional shear tests. *Can. Geotech. J.* **2011**, *39*, 22–31. [CrossRef]

8. Mazzuoli, M.; Bovolenta, R.; Berardi, R. Experimental Investigation on the Mechanical Contribution of Roots to the Shear Strength of a Sandy Soil. *Proced. Eng.* **2016**, *158*, 45–50. [CrossRef]

9. Bordoni, M.; Meisina, C.; Vercesi, A.; Bischetti, G.B.; Chiaradia, E.A.; Vergani, C.; Chersich, S.; Valentino, R.; Bittelli, M.; Comolli, R.; et al. Quantifying the contribution of grapevine roots to soil mechanical reinforcement in an area susceptible to shallow landslides. *Soil Tillage Res.* **2016**, *163*, 195–206. [CrossRef]

10. Bovolenta, R.; Mazzuoli, M.; Berardi, R. Soil bio-engineering techniques to protect slopes and prevent shallow landslides. *Riv. Ital. Geotec.* **2018**, *52*, 44–65.

11. Balzano, B.; Tarantino, A.; Ridley, A. Preliminary analysis on the impacts of the rhizosphere on occurrence of rainfall-induced shallow landslides. *Landslides* **2019**, *16*, 1885–1901.

12. Bittelli, M.; Valentino, R.; Salvatorelli, F.; Rossi Pisa, P. Monitoring soil-water and displacement conditions leading to landslide occurrence in partially saturated clays. *Geomorphology* **2012**, *173*, 161–173. [CrossRef]

13. Comegna, L.; Damiano, E.; Greco, R.; Guida, A.; Olivares, L.; Picarelli, L. Field hydrological monitoring of a sloping shallow pyroclastic deposit. *Can. Geotech. J.* **2016**, *53*, 1125–1137. [CrossRef]

14. Bordoni, M.; Valentino, R.; Meisina, C.; Bittelli, M.; Chersich, S. A simplified approach to assess the soil saturation degree and stability of a representative slope affected by shallow landslides in Oltrepò Pavese (Italy). *Geosciences* **2018**, *8*, 472. [CrossRef]

15. Bovolenta, R.; Passalacqua, R.; Federici, B.; Sguerso, D. Monitoring of Rain-Induced Landslides for the Territory Protection: The AD-VITAM Project. *Lect. Notes Civ. Eng.* **2020**, *40*, 138–147.

16. Bovolenta, R.; Passalacqua, R.; Federici, B.; Sguerso, D. LAMP-LAndslide Monitoring and Predicting for the analysis of landslide susceptibility triggered by rainfall events. In *Landslides and Engineered Slopes. Experience, Theory and Practice, Proceedings of the 12th International Symposium on Landslides, Napoli, Italy, 12–19 June 2016*; CRC Press, Taylor & Francis Group: London, UK, 2016; Volume 1, pp. 517–522.

17. Passalacqua, R.; Bovolenta, R.; Federici, B.; Balestrero, D. A physical model to assess landslide susceptibility on large areas: recent developments and next improvements. *Proced. Eng.* **2016**, *158*, 487–492. [CrossRef]

18. Passalacqua, R.; Bovolenta, R.; Federici, B. An integrated hydrological-geotechnical model in GIS for the analysis and prediction of large-scale landslides triggered by rainfall events. In *Engineering Geology for Society and Territory: Landslide Processes*; Springer: Berlin, Germany, 2015; Volume 2, pp. 1799–1803.

19. Passalacqua, R.; Bovolenta, R.; Spallarossa, D.; De Ferrari, R. Geophysical site characterization for a large landslide 3-D modelling. In *Geotechnical and Geophysical Site Characterization 4-Proceedings of the 4th International Conference on Site Characterization 4, ISC-4*; CRC Press, Taylor & Francis Group: London, UK, 2013; Volume 1, pp. 1765–1771.

20. Federici, B.; Bovolenta, R.; Passalacqua, R. From rainfall to slope instability: an automatic GIS procedure for susceptibility analyses over wide areas. *Geomat. Nat. Hazards Risk* **2015**, *6*, 454–472. [CrossRef]

21. *Hydrology, National Engineering Handbook, Supplement A, Section 4, Chapter 10*; USDA: Washington, DC, USA, 1956.

22. Bordoni, M.; Bittelli, M.; Valentino, R.; Chersich, S.; Persichillo, M.G.; Meisina, C. Soil Water Content Estimated by Support Vector Machine for the Assessment of Shallow Landslides Triggering: The Role of Antecedent Meteorological Conditions. *Environ. Model. Assess.* **2018**, *23*, 333–352. [CrossRef]

23. Coppola, L.; Reder, A.; Rianna, G.; Pagano, L. The Role of Cover Thickness in the Rainfall-Induced Landslides of Nocera Inferiore 2005. *Geosciences* **2020**, *10*, 228. [CrossRef]

24. Skempton, A.W.; DeLory, F.A. Stability of natural slopes in London Clay. In *Proceedings of the 4th International Conference on Soil Mechanics and Foundation Engineering*; Butterworths: London, UK, 1957; Volume 2, pp. 378–381.

25. Alvioli, M.; Baum, R.L. Parallelization of the TRIGRS model for rainfall-induced landslides using the message passing interface. *Environ. Model. Softw.* **2016**, *81*, 122–135. [CrossRef]

26. Baum, R.L.; Savage, W.Z.; Godt, J.W. *TRIGRS-A Fortran Program for Transient Rainfall Infiltration and Grid-Based Regional Slope-Stability Analysis: Open-File Report 02-424*; U.S. Geological Survey: Reston, VA, USA, 2002; p. 38.

27. Baum, R.L.; Savage, W.Z.; Godt, J.W. *TRIGRS-A Fortran Program for Transient Rainfall Infiltration and Grid-Based Regional Slope-Stability Analysis, Version 2.0: Open-File Report 2008-1159*; U.S. Geological Survey: Reston, VA, USA, 2008.

28. Montrasio, L.; Valentino, R. A model for triggering mechanisms of shallow landslides. *Nat. Hazards Earth Syst. Sci.* **2008**, *8*, 1149–1159.

29. Passalacqua, R.; Bovolenta, R. Landslides' susceptibility on large surfaces triggered by rain histories. In *Geotechnical Engineering for Infrastructure and Development, Proceedings of the XVI European Conference on Soil Mechanics and Geotechnical Engineering, Edinburgh, UK, 13–17 September 2015*; British Geotechnical Association: London, UK, 2015; Volume 4, pp. 1831–1836.

30. Chirico, G.B.; Borga, M.; Tarolli, P.; Rigon, R.; Preti, F. Role of vegetation on slope stability under transient unsaturated conditions. *Proced. Environ. Sci.* **2013**, *19*, 932–941. [CrossRef]

31. Sguerso, D.; Labbouz, L.; Walpersdorf, A. 14 years of GPS tropospheric delays in the French–Italian border region: comparisons and first application in a case study. *Appl. Geomat.* **2015**, *8*, 1–13. [CrossRef]

32. Ferrando, I.; Federici, B.; Sguerso, D. 2D PWV monitoring of a wide and orographically complex area with a low dense GNSS network. *Earth Planets Space* **2018**, *70*, 1–21. [CrossRef]

33. Bishop, A.W. The principle of effective stress. *Tek. Ukebl.* **1959**, *106*, 859–863.

34. Fredlund, D.G.; Morgenstern, N.R.; Widger, R.A. The shear strength of unsaturated soils. *Can. Geotech. J.* **1978**, *15*, 313–321. [CrossRef]

35. Lu, N.; Godt, J. Infinite slope stability under steady unsaturated seepage conditions. *Water Resour. Res.* **2008**, *44*, 1–13. [CrossRef]

36. Campora, M.; Palla, A.; Gnecco, I.; Bovolenta, R.; Passalacqua, R. The laboratory calibration of a soil moisture capacitance probe in sandy soils. *Soil Water Res.* **2019**, *15*, 75–84. [CrossRef]
37. *Calibration Manual for Sentek Soil Moisture Sensors, Version 2.0*; Sentek Sensor Technologies: Stepney, Australia, 2011.
38. *Retriever & Pup Wireless Network: Product Manual*; Spectrum Techologies, Inc.: Thayer Court Aurora, IL, USA, 2015.
39. *Soil Sensor Reader: Product Manual*; Spectrum Techologies, Inc.: Thayer Court Aurora, IL, USA, 2015.
40. *WaterScout SM100 Soil Moisture Sensor: Product Manual*; Spectrum Techologies, Inc.: Thayer Court Aurora, IL, USA, 2015.
41. Pineda-Jaimes, J.A.; Murillo-Feo, C.A.; Colmenares, J.E. Characterization of pavement pathologies associated with the action of plant species in a road at western Sabana de Bogota. *Épsilon* **2015**, *25*, 39–68.
42. Bovolenta, R.; Federici, B.; Berardi, R.; Passalacqua, R.; Marzocchi, R.; Sguerso, D. Geomatics in support of geotechnics in landslide forecasting, analysis and slope stabilization. *Geoing. Ambient. Miner.* **2017**, *151*, 57–62.

Article

Evaluation of Unsaturated Soil Properties for a Debris-Flow Simulation

Francesco Castelli, Valentina Lentini * and Alessandra Di Venti

Faculty of Engineering and Architecture, University of Enna "Kore", 94100 Enna, Italy;
francesco.castelli@unikore.it (F.C.); alessandra.diventi@unikore.it (A.D.V.)
* Correspondence: valentina.lentini@unikore.it; Tel.: +39-0935-536351

Abstract: Fast-moving landslides (i.e., debris/earth flows) are often caused by heavy rainfall occurring in small areas, and are not predictable. On the other hand, innovative methods for geomechanical characterization, numerical analysis, and modeling are required to attempt to reproduce a given debris/earth flow event. As our capability to reproduce very complex phenomena increases, we can improve our prevention approaches. In this paper, a debris flow event that occurred in the Enna area (Sicily) is described. Starting from the study of the geological framework and the historical background, this research focused on the causes that triggered the landslide. In situ and laboratory tests, including geophysical investigations and triaxial tests in unsaturated conditions, were carried out to investigate the factors affecting the dynamics of the event. This study gives us better knowledge of the mechanical and hydraulic properties that can be used to model these events, to assess the most appropriate strategies for the prevention and mitigation of related risks.

Keywords: debris flow; in situ characterization; triaxial tests; unsaturated conditions

Citation: Castelli, F.; Lentini, V.; Venti, A.D. Evaluation of Unsaturated Soil Properties for a Debris-Flow Simulation. *Geosciences* **2021**, *11*, 64. https://doi.org/10.3390/geosciences 11020064

Academic Editors: Roberto Valentino and Jesus Martinez-Frias
Received: 27 December 2020
Accepted: 27 January 2021
Published: 31 January 2021

Publisher's Note: MDPI stays neutral with regard to jurisdictional claims in published maps and institutional affiliations.

1. Introduction

Fast-moving landslides (i.e., debris/earth flows) are among the natural phenomena which produce severe damage and fatalities. The high velocity that the flow mass can reach during propagation, due to the characteristics of moving material, allow an extended area to be rapidly covered, so the consequences of a "rapid landslide" impact are pronounced when it occurs near structures or a lifeline. Therefore, in the last few decades, many efforts have been made to investigate the factors affecting the dynamics of these events, and to develop models able to simulate the flow propagation, with the aim of producing landslide maps and developing appropriate strategies for the prevention and mitigation of related risks.

To achieve reliable predictions, information about the three components of risk (hazards, exposition, and vulnerability), and geological and geotechnical data, are strictly necessary. With this aim, here, with the traditional in situ and laboratory characterization, innovative procedures are introduced to investigate soil behavior, such as, for example, triaxial laboratory tests with suction control [1]. Numerical models, likewise, have been developed to simulate landslide flow propagation and run-out [2–9].

From a theoretical point of view, debris flows are very peculiar. At first approximation, they can be classified as suspensions—mixtures of fluid and grains. The fact that the grain size is not uniform complicates their understanding: the smallest grains are microscopic clay particles, while the biggest ones are large-sized granular soil. The interaction between the grains and fluid also takes different forms depending on the grain size.

The dynamics of microscopic grains are dominated by electrostatic forces, while big grains are driven by hydrodynamic interactions with the fluid.

Numerical models developed for simulating debris/earth flows generally refer to hyper-concentrated flows, which differ in their adopted rheological schemes. In particular, they can be separated into single-phase models and two-phase models.

Single-phase models assume that a debris flow acts as a homogeneous Bingham fluid, composed of a mixture of water and sediment. From a rheological point of view, such a mixture can be described by Herschel–Bulkley [10], or in even more complex models such as, for example, that described by a quadratic law that assumes that the total friction stresses can be divided into different terms—yield stresses, viscosity stresses, and turbulent-dispersive stresses—all of them being functions of the sediment concentration in the mixture.

The hypothesis of the Binghamian nature of the fluid is necessary in order to simulate the arrest of the flow. In the two-phase models, the exchange of mass between the erodible bed and the flow is taken into account as well. The fundamentals of such models were first developed by Bagnold [11], and then applied to the debris flows by Takahashi [12].

In these models, the solid concentration is an unknown variable which influences the global behavior of the flow, which can be properly accounted for by the model itself. In particular, it has been noted that the first type of model is more suitable for cases characterized by fine sediments, when the viscous shear rate is high. The second type of model is more suitable in cases in which the viscosity of the interstitial fluid is negligible and the solid fraction is composed of coarser material, so that the inertial shear rate acts predominately due to the collisions between gravels.

The choice of the rheology and related rheological parameter values is a fundamental aspect of numerical simulation; therefore, to improve knowledge of the rheological properties of the debris flow (e.g., the concentration of solid material, evolutionary characteristics of viscous water/soil mass movement, mixture density, viscosity, etc.), direct comparisons between numerical predictions and results deduced by case histories or physical model simulations can be used.

On the other hand, evaluation of the hydromechanical properties of the unsaturated soils by innovative laboratory tests is another important aspect of understanding debris/earth flows.

Difficulties in numerical modeling of this heterogeneous moving mass are mainly linked to the simulation of the complex behavior assumed by the mass during propagation. As stressed by Iverson et al. [13], the debris flow during propagation does not behave with a fixed rheology, since it changes its rheological characteristics in space and time.

To derive information for the calibration of the model parameters, the results of an experimental study and the back-analysis of a debris flow undertaken in the Enna municipality (Sicily, Italy) on February 2014 are reported and discussed here. The numerical simulations were carried out using a pseudo-2D model developed for debris flow modeling that adopts depth-integrated flow equations.

2. The Case Study

Due to its nature and the complex morphological, geological, and lithotechnical conditions, as well as the particularly non-homogeneous and complicated hydrogeological aspects, the Sicilian territory is characterized by various types of landslides (mainly debris/earth flows) and floods. Because of the progressive neglect of mountainous areas and a lack maintenance of the slopes and hydraulic structures used to regulate rainwater, events related to hydrogeological risks can cause significant damage. The study area is located close to the municipality of Enna, in the middle of Sicily, and is characterized by a morphology with very high hills and catchment areas of small to moderate extensions (about 0.3 km^2).

During the night between the 1st and the 2nd of February 2014, heavy rainfall struck the area (about 150 mm of rainfall in less than 1 h), causing severe damage and interrupting the main infrastructure connecting to the city at the highway "A19 Palermo-Catania" (Figure 1a). Figure 1b shows the landslide path from the source area to the accumulation zone.

Figure 1. Enna event (2014): (**a**) soil deposited on the road, (**b**) damage that occurred to infrastructure connecting the city at the highway "A19 Palermo-Catania".

A basin ends at the base of the slope, and represents the deposition zone of the rapid flow phenomena, including for the event that occurred in February 2014 (Figure 2).

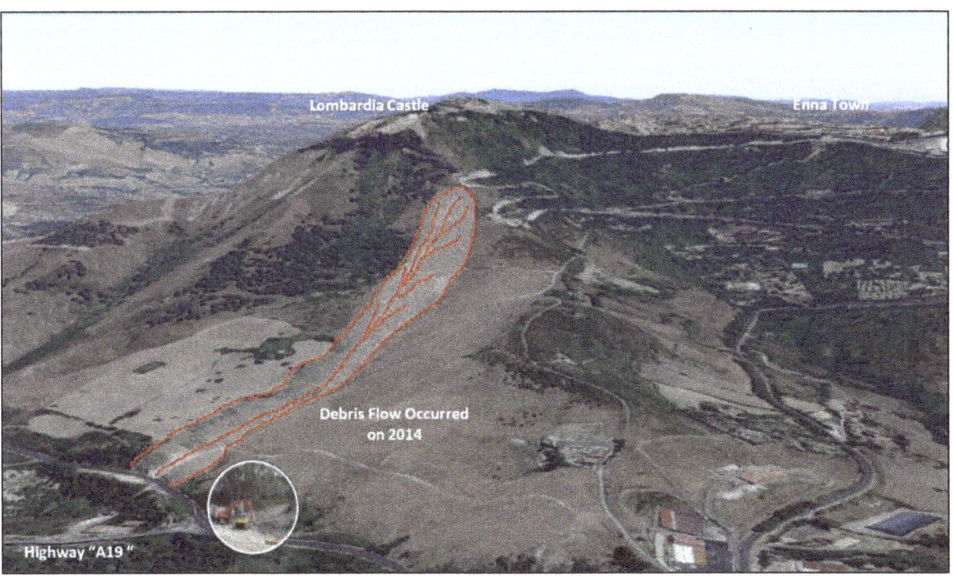

Figure 2. View of the debris flow path in February 2014 in Enna (Italy).

2.1. The 2014 Enna Debris Flow

Enna is the least populated Sicilian province. Its altitude ranges from 58 to 1561 m a.s.l., with a morphology of high hills with slope angles ranging from 15° to 60° (Figure 3). The landscape features a high number of streams and lakes, mostly artificial and used for irrigation purposes.

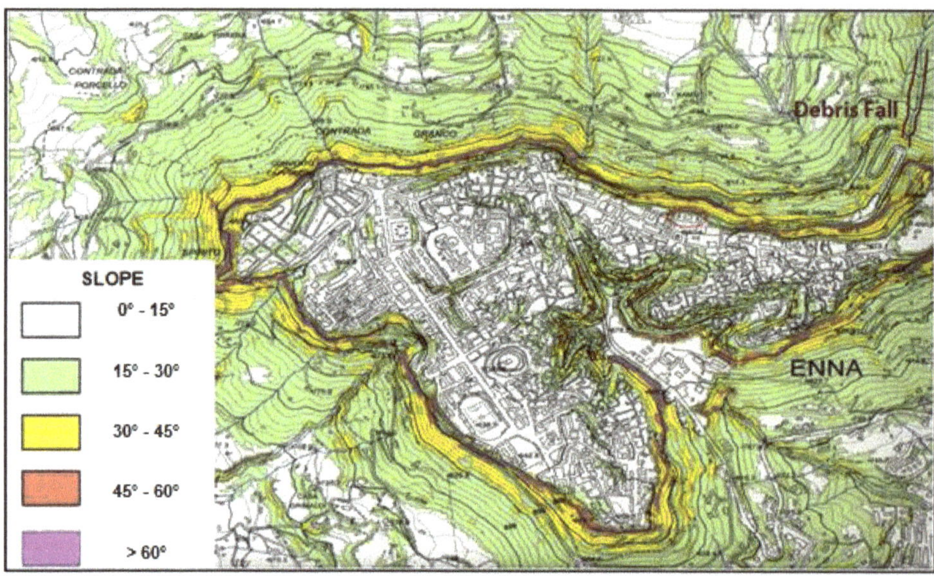

Figure 3. Slope map of the Municipality of Enna (Italy).

The city of Enna is located in a hilly area, and from a geological point of view (Figure 3), is made up of Numidian Flysch of lower Oligocene age, marly and sandy brown clay of medium Miocene age, and river alluvium of Holocene age [14]. The climate is characterized by an average daily temperature of about 14 °C, with fluctuations between day and night of about 15 °C. Usually, short and intense rainstorms occur between October and March.

The event of 2nd February 2014 occurred in an area characterized by a catchment with a small extension of 0.158 km^2. This area is characterized by a high urban density with narrow roads, that during the event became the bed of the runoff flows.

The Enna area is characterized by the following main lithostratigraphic units [15]: "Trubi", Enna marls, calcarenites, gray-dark brown "brecciated" clays, and gray-blue clays (Figure 4).

Superficial gravitational movements are triggered at the level of the altered blue-gray clayey marls, having thicknesses varying from 2.0 m up to about 4.0 m. These phenomena mainly affect the upper part of the geological formation, especially during intense rainy phenomena.

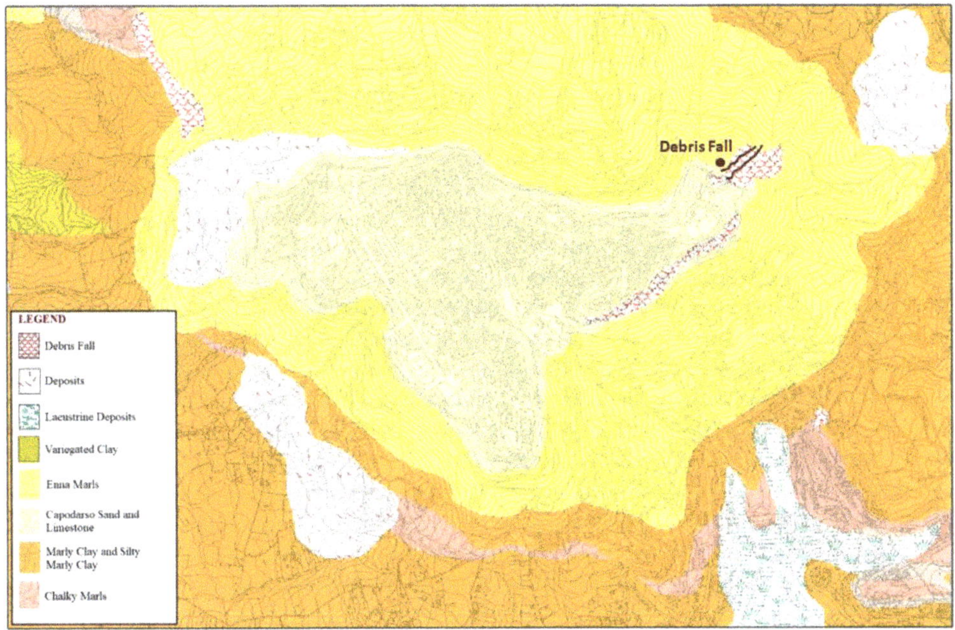

Figure 4. Geological map of the Municipality of Enna (Italy).

2.2. Site Characterization and Geotechnical Soil Properties Evaluation

The test site was accurately investigated by in situ surveys and laboratory tests. Boreholes, electrical tomographies (Figure 5), and standard and cone penetration tests were carried out in areas close to the landslide event (Figure 6). The boreholes, in particular, showed that the debris flow mainly consisted of cohesionless silty sand materials (Figure 7).

Laboratory tests on soil samples were carried out for evaluation of their geotechnical properties. The main geotechnical parameters, derived by laboratory tests performed on soil samples retrieved from a survey carried out in the area on the debris flow event, are reported in Table 1. It shows the physical parameters expressed in terms of soil unit weight γ, water content w_n, Atterberg limits w_L and w_p, degree of saturation S_r, and strength properties derived by means of direct shear tests and standard unconsolidated undrained triaxial tests, in terms of friction angle ϕ', cohesion c', and undrained cohesion c_u, respectively.

Table 1. Geotechnical soil parameters.

Sample	Depth (m)	γ (kN/m^3)	w_n (%)	w_L (%)	w_p (%)	S_r (%)	c' (kPa)	ϕ' (°)	c_u (kPa)
S1/C1	2.10	18.45	22.2	48	26	90	0.29	17	-
S1/C2	2.90	19.40	11.2	42	19	97	0.41	24	211
S1/C3	12.90	20.61	17.3	39	20	95	-	-	-

The geotechnical soil parameters reported in Table 1 show that the samples recovered in the area of the debris flow are almost saturated. Evaluation of the hydromechanical properties of the unsaturated soils was an important aspect of this study on this landslide.

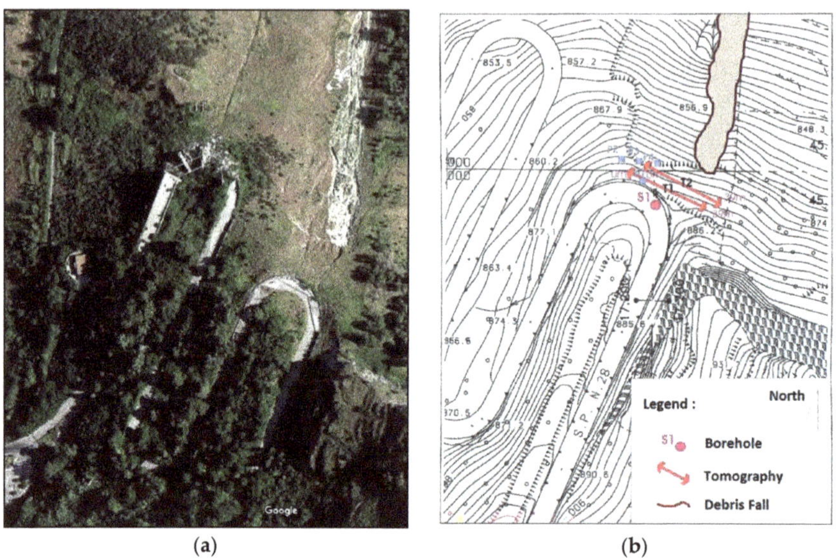

Figure 5. The test site: (**a**) plan view, and (**b**) layout of the test site with borehole and tomography locations.

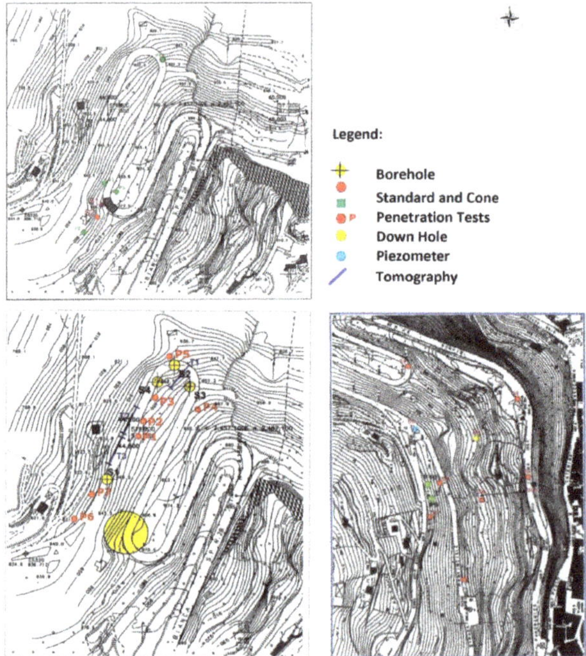

Figure 6. Plan view of the field investigations carried out in the area.

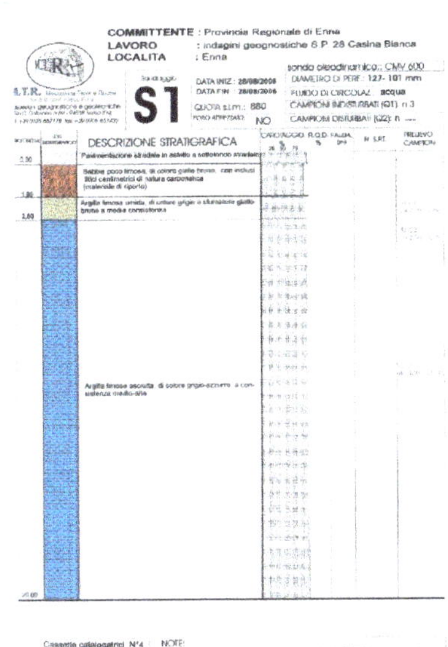

Figure 7. The test site: soil profile derived by the borehole.

In Figures 8 and 9, the results of two tomographies carried out very close to the run-out area and along the path of the debris flow, respectively, are shown. In the first case (Figure 8), the results are expressed in terms of primary wave seismic velocities V_p (m/s). In the second case, the results are expressed by the electrical resistivity model (Figure 9a) and the chargeability model (Figure 9b), obtained using the electrical resistivity tomography (ERT) method. In this last case, the high correlation of electrical resistivity to soil moisture can be used to detect anomalies (Figure 9b), which can be related to an unsaturated material, potentially subjected to instability phenomena [16].

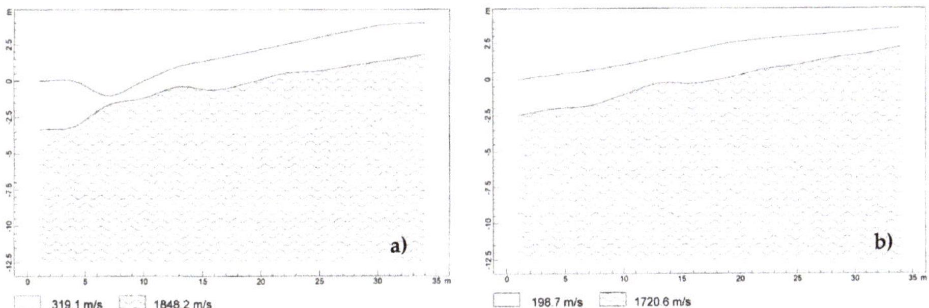

Figure 8. Primary wave seismic velocities V_p derived by (**a**) tomography T1 and (**b**) tomography T2.

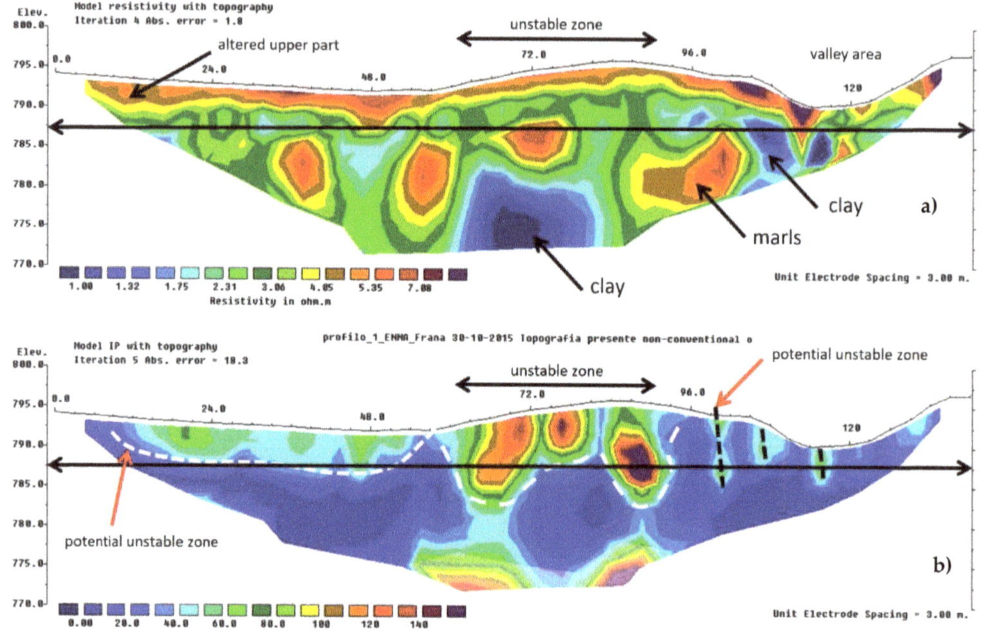

Figure 9. Resistivity model (**a**) and chargeability model (**b**) of the soil along the path of the debris flow derived by electrical tomography.

2.3. Laboratory Triaxial Tests in Unsaturated Conditions

Unsaturated soils with pore water pressures that are negative relative to atmospheric conditions are widespread, especially at shallow depths from the surface. The shear strength parameters of these soils are crucial for stability analyses of surface soils, such as analyses of slopes against slope failures and landslides. Saturated soil mechanics are a complex subject, consisting of the solid phase form and its interaction with the aqueous phase.

When the soil dries such that there is also an air phase, it becomes unsaturated and its behavior is far more complex because of the interface between the air and the water, and the volumetric response of the air under changing conditions of pressure and temperature.

Additionally, the pressure in the water phase becomes negative (tensile), and measuring such pressures has, until recently, been fraught with problems.

In this study, laboratory tests were carried out to investigate the influence of matric suction on the shear strength and mechanical behavior of soil. The axis translation technique was applied, and a double-walled triaxial cell was used to measure the soil matric suction and variation of pore air volume, respectively. It can be deduced that matric suction influences the shear strength of unsaturated soils, and their failure mode in the final state of loading. Suction increases with an increase in axial strain during monotonic shearing.

2.3.1. Mechanical Behavior of Unsaturated Soil

Engineering tests show that there are significant differences between the mechanical behavior of unsaturated soils and completely saturated or completely dry soils. Normally, the theory of soil mechanics considers the soil as completely saturated or dry. However, it has been discovered that there are unsaturated areas of the soil, where gaps between the soil particles are filled with a mixture of air and water. Often in practice, these areas have been underestimated when assuming that the soil is completely saturated or completely dry.

Three phases characterize the unsaturated soil: solid particles, water, and air. The mechanical and hydraulic behavior and the tensional state are influenced by the presence of these two fluids inside the pores (one of which is compressible), and the interaction between them and with the solid skeleton. The presence of the three phases makes it necessary to take into account the differences in air and water pressures, phase compressions and their interactions, and chemical effects. In fact, the contractile skin between the fluid phases causes a surface tension effect that gives the shape of the characteristic aggregate structure of fine-grained soils.

The physical quantity that characterizes unsaturated soils is matric suction, and it is defined as the difference between pore air pressure and pore water pressure. The resistance of unsaturated soils has been formulated in terms of two independent stress variables: the net normal stress and matric suction [17].

Triaxial laboratory tests are very useful in suction control, using the axis translation technique, for the evaluation of the main features (i.e., the deviator stress, the pore air pressure and pore water pressure, the axisymmetrical volume change, and the air/water volume changes). The water content of unsaturated soil has a direct influence on soil suction, and so the relationship between water content and suction is sought in the form of a soil water characteristic curve (SWCC).

2.3.2. Soil–Water Characteristic Curve (SWCC)

In a Soil–Water Characteristic Curve (SWCC), suction can be expressed in terms of matric suction $(u_a - u_w)$ or total suction Ψ. However, in practice, the total suction can be assumed to be equal to the matrix (for high values of suction up to 3000 kPa), neglecting osmotic suction.

The total suction corresponding to a water content of zero (close to 10^6 kPa) seems to be substantially the same for all types of soil [18]. Figure 10 shows the typical SWCC diagrams of volumetric water content θ_w against suction s = $(u_a - u_w)$ for a range of soils from Holland [19].

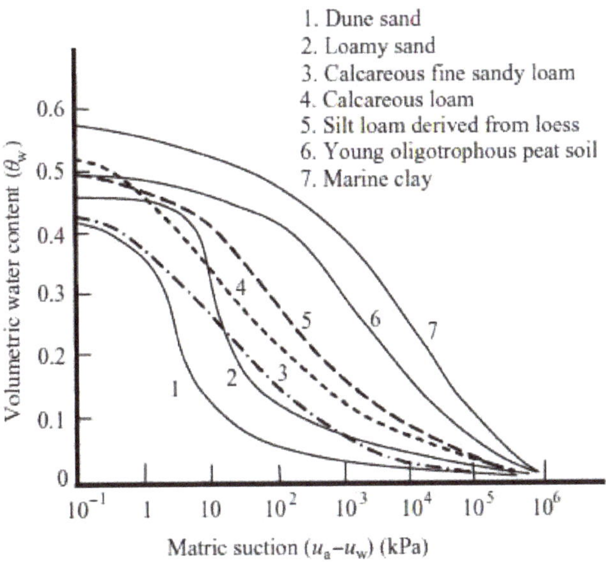

Figure 10. Soil–Water Characteristic Curves (SWCC) for some Dutch soils (after Koorevaar et al. [19]).

The behavior of saturated soils is described by the principle of effective stress [20], but it would be a mistake to apply it to unsaturated soils characterized by the presence of air, water, and air–water interfaces in the voids.

Fredlund et al. [21] formulated the shear strength equation for an unsaturated soil as:

$$\tau_f = c' + (\sigma - u_a)_f \tan \varphi' + (u_a - u_w)_f \tan \varphi^b \tag{1}$$

where τ_f = shear stress at failure, c' = apparent effective cohesion, $(\sigma - u_a)_f = (\sigma_{net})_f$ = the net stress at failure, σ = total normal stress, u_a = pore air pressure, φ^b = angle indicating the rate of increase in shear strength relative to the matric suction at failure, and u_w = pore water pressure.

Equation (1) can be reduced to an effective stress equation as follows:

$$\tau_f = c' + [(\sigma - u_a)_f + \chi_f (u_a - u_w)_f]\tan \varphi' \tag{2}$$

where χ_f = the effective stress parameter, varying between 0 and 1.

Vanapalli et al. [18] examined the relationship between the degree of saturation and χ_f, based on the shear strength results reported by Escario et al. [22]. The following expressions for χ_f are in good agreement with the shear strength for matric suctions ranging between 0 and 1500 kPa.

$$\chi_f = S^k = \left(\frac{\theta}{\theta_s} \right)^k \tag{3}$$

$$\chi_f = \frac{S - S_r}{1 - S_r} = \frac{\theta_w - \theta_r}{\theta_s - \theta_r} \tag{4}$$

where S = degree of saturation, θ_w = volumetric water content, θ_s = saturated volumetric water content, κ = parameter optimized to obtain the best fit between measured and predicted values, θ_r = residual volumetric water content, and S_r = residual degree of saturation.

Thu et al. [23] found from constant water content triaxial testing of compacted silt that φ^b was not linearly related to matric suction. The value of φ^b at low suction (soil remaining saturated) equaled the effective stress friction angle φ'.

2.3.3. Axis Translation Technique

The axis translation technique was devised by Hilf [24] for the control of suction in unsaturated soils during triaxial laboratory tests [25]. This technique is based on the principle of obtaining the measurement of positive values of pore water pressure in unsaturated soils by increasing the pore air pressure and total stress, and preventing cavitation. The difference between pore air pressure and the pore water pressure is suction.

The triaxial cell is formed by a double cell that has the same pressure inside and outside the internal cell (Figure 11). There will be no expansion of the internal cell, which allows the use of a transducer of volume change in the cell pressure line. During the test, it is necessary to measure the volume of water leaving the sample, and the variation of the total sample volume.

A saturated high-air entry ceramic disc is used to prevent air from entering into the pore water measuring system (ceramic discs are produced up to an air inlet value of 1500 kPa). Complete saturation of the disc is essential before its use in a test. Wheeler and Sivakumar [25] recommended a procedure for this purpose.

In the measuring system, the air entry value of the ceramic disc must be higher than the matric suction to be measured, to avoid pore water going into the system. This increases if the pore size of the disc decreases, controlling the curvature radii of the air–water menisci at the filter boundary.

This technique has two disadvantages: first, in situ stress conditions cannot be replicated in laboratory tests; second, there is a risk of air diffusion through the high air entry filter into the pore water pressure system, producing air bubbles and influencing the acquired measurements.

Different types of stress path tests could be carried out, such as constant suction, constant net stress, constant volume tests, constant rates of strain, or fully drained or undrained strain-controlled tests.

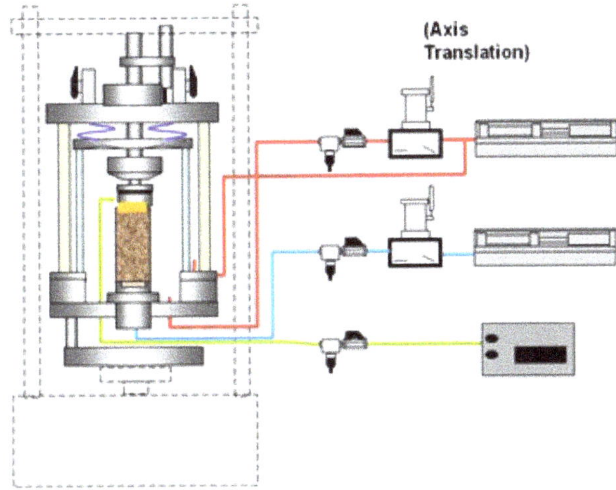

Figure 11. Axis translation in the triaxial apparatus.

2.4. Laboratory Testing Program

A better understanding of the behavior of unsaturated soils has been obtained from laboratory and in situ studies. In this field, the development of suction tests was a major breakthrough that has radically changed testing methodologies.

Today, perhaps the greatest advances have been reached with the implementation of more complete constitutive models for unsaturated soils in numerical analyses. Laboratory test data are still required to supply representative parameters for the constitutive models, and to provide frameworks of behavior against which the implemented models can be validated.

Most existing constitutive models are based on laboratory testing programs that have been achieved using the axis-translation technique, where pore pressures are positive. In this work, the unsaturated stress–strain behaviors of soil specimens were investigated by triaxial tests carried out in the Geotechnical Laboratory of the University of Enna "Kore" (Sicily, Italy).

Figure 12 shows a sketch of the testing apparatus used in this study. During the triaxial tests, the following stages were performed: saturation, isotropic consolidation, SWCC, and monotonic shear using the triaxial machine.

The physical properties of the soil specimens are given in Table 2. The soil sample used in this study was a sandy clayey silt (Figure 13), with a soil particle density of 24.76 kN/m^3.

The specimen was packaged through a filter paper and porous stone at the head and high air entry stone (Haes) at the base, and wrapped in a membrane and o-rings. First, the Haes disk was saturated for 24 h with a pressure of 800 kPa before starting the test.

In the SWCC stage, the air pressure at the top of the soil sample was increased and decreased in steps of 20 kPa above the water pressure of the pores, recording the different suction values for the construction of the Soil–Water Characteristic Curves (SWCC). Initially, the matric suctions (plateau of the SWCC) were increased to about 100 kPa (Table 3). Then, in the monotonic shear stage, a rate of strain of 0.30%/hr was calculated, and an axial strain ε of 20% was established. In the final state of loading, the failure mode shown in Figure 14 was observed.

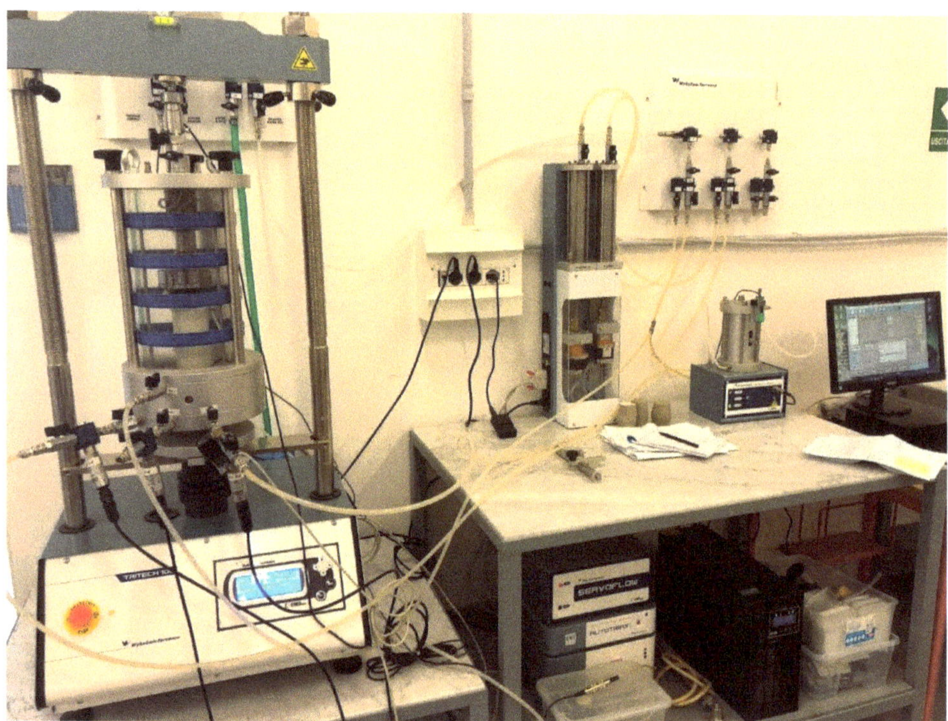

Figure 12. Triaxial machine for unsaturated soil in the Geotechnical Laboratory of the University of Enna "Kore".

Table 2. Specimen features and physical properties of the soil.

Specimen Features				
Length	100.22 mm	Area	1967.4	mm^2
Diameter	50.05 mm	Volume	197.20	cm^3
Wet Mass	381.25 g	Volume of Voids	87.82	cm^3
Particle Density	2.52 Mg/m^3	Voids Ratio	0.803	
Dry Mass	291.84 g	Bulk Density	1.93	Mg/m^3
Initial Moisture Content				
Mass of Wet Soil + Tin	111.06 g	Mass of Tin	18.27	g
Mass of Dry Soil + Tin	89.30 g	Moisture Content	30.63	%

Table 3. Suction steps in the initial SWCC stage.

Step Type	u_a (kPa)	$(u_a - u_w)$ (kPa)
Initial	87	–3
AP Increment	106	17
AP Increment	134	45
AP Increment	145	56
AP Increment	171	82
AP Increment	184	95

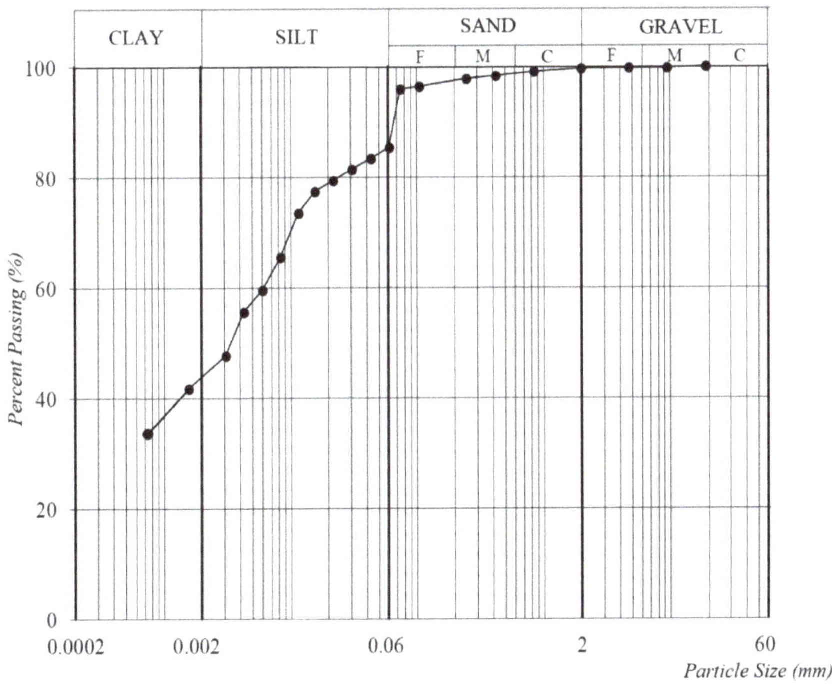

Figure 13. Grain size distribution curve of the soil specimen.

Figure 14. Failure mode of unsaturated specimens at the lowest suction value.

In carrying out the test, it was possible to analyze the behavior of the specimen subjected to different load steps until the established strain value was reached.

In Figure 15a,b, the relationship between soil suction and volumetric water content, and the relationship between deviatoric stress and axial strain for an unsaturated specimen are reported, respectively. Finally, in Figure 16a,b, the volumetric strain versus axial strain relationship and the stress-path for the unsaturated specimen on the (s'; t) plane are reported, respectively, with $s' = (\sigma'_1 + \sigma'_3)/2$ and t the maximum shear stress $(\sigma_1 - \sigma_3)/2$ (critical state variable for the unsaturated soil).

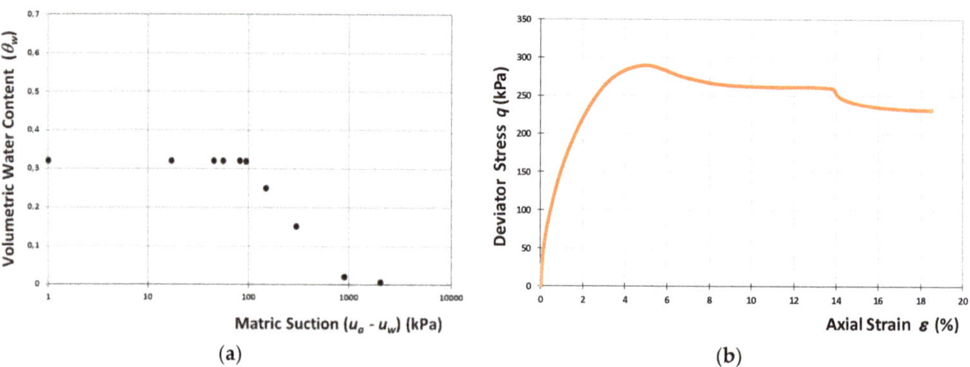

(a) **(b)**

Figure 15. Matric suction–volumetric water content (**a**,**b**) deviator stress–axial strain relationships.

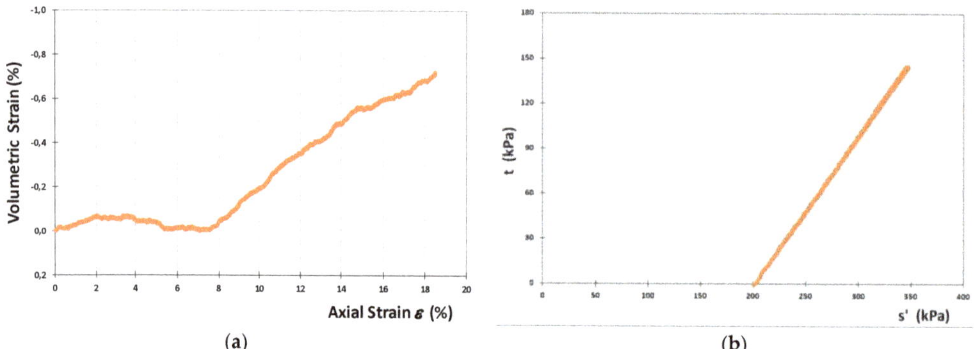

(a) **(b)**

Figure 16. Volumetric strain–axial strain relationship (**a**,**b**) effective stress-path.

3. Numerical Modeling

The aim of the proposed study was to investigate rainfall-induced landslides on partially saturated soil slopes, by modeling the February 2014 debris flow that occurred in the Enna municipality (Sicily, Italy). The specific objectives of this study were: (1) the identification of geomorphological features and the spatial model of the landslide, (2) integrated analysis through laboratory tests and field tests and, finally, (3) discussion of the results of the numerical analysis.

Propagation models applicable to hyper-concentrated flows differ for the adopted rheological schemes. Simplified single-phase models find an "equivalent" fluid, whose rheological properties are such that the bulk behavior of the flowing body can simulate the expected bulk behavior of the real landslide. Generally, the debris flow is considered as a homogeneous Bingham fluid, composed of a mixture of water and sediment. The hypothesis of the Binghamian nature of the fluid is necessary in order to simulate the arrest of the flow. The rheological characteristics can vary depending on the lithology, grain-size distribution, and sediment concentration.

To check the possibility of numerically simulating the debris flow event, the FLO-2D code [26], which is not fully two-dimensional, based on a monophasic Bingham scheme and modeled through the quadratic rheological law developed by O'Brien and Julien [27], was used.

This code is widely used, even though the setting of rheological parameter values for prediction purposes is still problematic. The presented numerical analysis, in particular, intended to investigate: (i) the capability of the code to simulate the dynamics of the debris flow (propagation and deposition) and (ii) the influence of the associated rheological parameters on the numerical results, calibrated from the soil parameters evaluated by in situ and laboratory tests, including non-conventional triaxial tests in unsaturated conditions.

3.1. The FLO-2D Model

FLO-2D is a commercial computer code developed by O'Brien [26], and adopted for debris flow phenomena modeling and delineating flood hazards. It is a pseudo-2D model in space that adopts depth-integrated flow equations.

In the FLO-2D model, hyper-concentrated sediment flows are simulated considering the flow as a homogeneous (monophase) non-linear Bingham fluid, based on an empirical quadratic rheological relationship developed by O'Brien and Julien [27].

In particular, the basic equations implemented in the model consist mainly of the well-known continuity equation and equation of motion, according to a total friction slope S_f expressed as:

$$S_f = \frac{\tau_B}{\rho g h} + \frac{k \mu_B V}{8 \rho g h^2} + \frac{n^2 V^2}{h^{\frac{4}{3}}} \tag{5}$$

where h is the flow depth, V is the depth-averaged velocity, g is the gravitational acceleration, τ_B is the Bingham yield stress, ρ is the mixture density, k is the laminar flow resistance coefficient, μ_B is the Bingham viscosity, and n is the pseudo-Manning resistance coefficient, which accounts for both turbulent boundary friction and internal collisional stresses, expressed as:

$$n = 0.538 \, n_t \, e^{\, 6.0896 C_v} \tag{6}$$

The yield stress τ_B, the dynamic viscosity μ_B, and the resistance coefficient n are influenced by the sediment concentration, and these parameters can be described by the following equations [13]:

$$\tau_B = \alpha_1 e^{\beta_1 C_v} \tag{7}$$

$$\mu_B = \alpha_2 e^{\beta_2 C_v} \tag{8}$$

where C_v is the volumetric concentration and α_1, α_2, β_1, and β_2 are the empirical coefficients defined by laboratory tests performed by O'Brien and Julien [27], while n_t is the turbulent n value.

More detailed information about the numerical scheme and the general constitutive fluid equations adopted can be found in O'Brien [26].

3.2. Back Analysis Procedure

Several applications of the FLO-2D model can be found in the literature, which mainly differ in relation to the sediment characteristics, and hence, in the adopted rheological parameters [28–31]. Various comparisons with other methodologies [32,33] have demonstrated that FLO-2D, if appropriately calibrated, represents a useful tool for predicting the behavior of future landslides of the same type and in similar settings.

In this case, to construct the DTM for modeling the debris flow in Enna (Figure 17), a grid system with a cell size of 2.0 × 2.0 m was implemented by the FLO-2D code [34].

Figure 17. Location of the triggering (1) and deposition (2) zone of the debris flow.

The hydrological input reported in Figure 18a,b was applied at the upstream section of the catchment basin, where the triggering was observed [35–38]. The rainfall from the ground surface is a key factor in slope instability. Many debris flows begin as shallow landslides induced by direct infiltration of intense rainfall and runoff into cover materials [39–45]. In FLO-2D, the precipitation is an input and it starts after triggering to estimate the propagation velocities and the mobilized volumes.

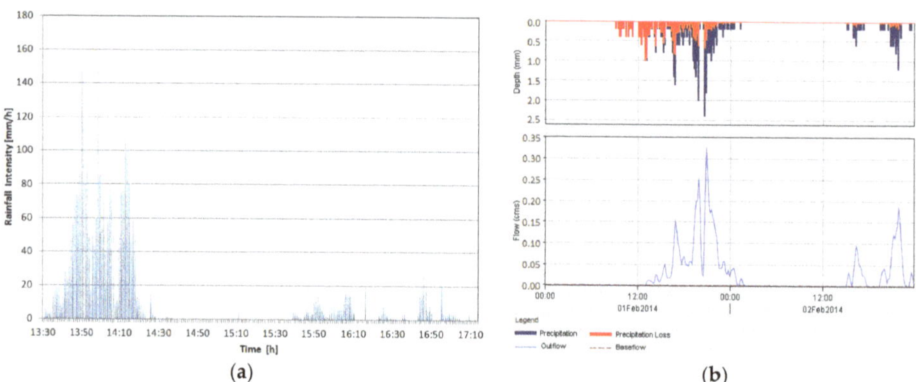

(a)   (b)

Figure 18. (a) Recorded rainfall intensity, and (b) hydrological input used in the FLO-2D simulation.

The discharge rate value of the debris flow in the upper part of the catchment basin was calculated according to the expression proposed by Armanini et al. [46].

The following values of the empirical coefficients defined by the laboratory tests performed by O'Brien and Julien [27], $\alpha_1 = 0.0060$, $\beta_1 = 19.9$, $\alpha_2 = 0.0007$, and $\beta_2 = 29.8$, were chosen for the numerical simulation, with the aim of selecting those corresponding to similar characteristics of the area from morphological and lithological points of view [31].

The presence of the road has been taken into consideration. Assuming a value of the Manning resistance coefficient equal to $n = 0.04$, the debris flow concentration was chosen equal to 0.57, while the slope angle was $\alpha = 20°$.

More detailed information about the numerical scheme and the general constitutive fluid equations adopted can be found in O'Brien [34].

According to Armanini et al. [47], a relationship can be found between the friction angle of the soil and the debris flow concentration. With this aim, in the future, triaxial tests will be carried out to investigate this possibility, to relate the values of the coefficients chosen in the numerical simulation to the strength parameters of the unsaturated soil.

3.3. Numerical Simulation

A reconstruction of the inundated area (Figure 19a) was obtained as the output of the simulations performed by FLO-2D [34]. The computed maximum flow depths during the event are presented in Figure 19b. The highest predicted flow depth is about 4.0 m. Figure 19c represents the final flow depths. The highest value of the predicted final flow depth is about 1.4 m.

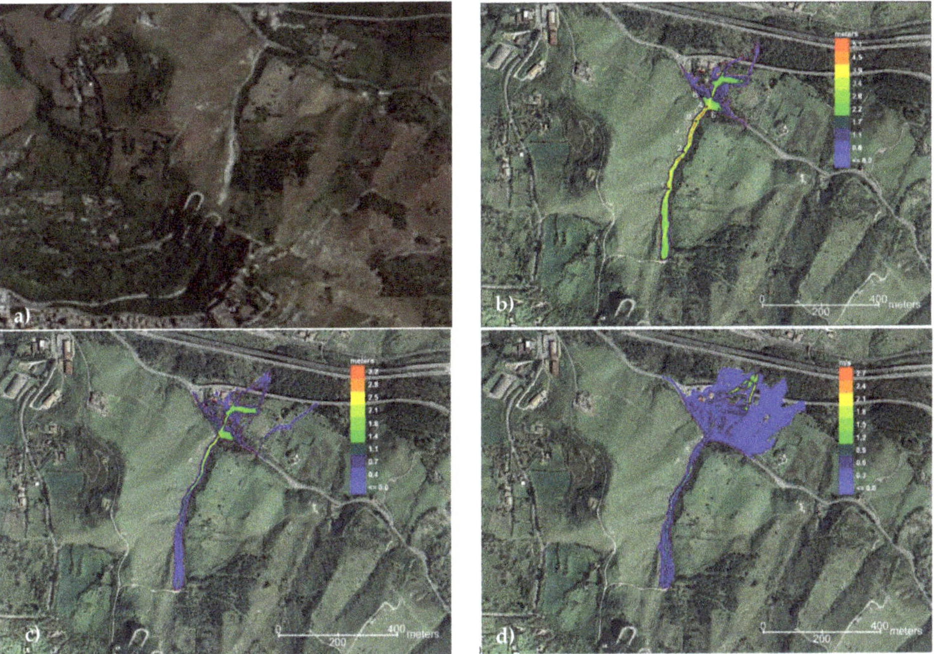

Figure 19. Scenarios simulated with FLO-2D for (**a**) the Enna debris flow: (**b**) maximum flow depth, (**c**) final flow depth and, (**d**) maximum velocities.

Finally, the predicted maximum velocities are shown in Figure 19d. It is possible to observe that the maximum velocities were registered corresponding to the upper part of the basins, where the slope is the highest, with values ranging from 1 to 2 m/s.

The predicted values are, in general, in good agreement with those observed. This is supported by the comparison between the computed volumes and the measurement of the deposited sediment after the event, resulting from the surveys of Regional Civil Protection [38].

Several numerical simulations have been performed, attributing different concentration values to the mixture, and it can be deduced that the simulations carried out by using higher concentration values for the incoming hydrograph led to results that differ a lot from reality, as deposit height is higher and inundated areas are reduced compared to in reality. When decreasing the concentration value, there is instead a considerable increase in the area affected by the debris flow, with a decrease in flow height due to a greater fluidity of the mixture.

From this point of view, the present study gives us better knowledge of the hydraulic parameters that can be used for other models in areas with similar soil composition, with the aim to investigate the global dynamics of the event.

These results can be applied to risk calculations, reproducing in thematic maps by different risk levels the distribution of the flow mass on the propagation path, its intensity, and the zone where the elements will experience an impact.

4. Concluding Remarks

Debris flows are natural phenomena which often cause severe damage and fatalities. Therefore, in the last few decades, many efforts have been made to develop models able to numerically simulate debris flow propagation, with the aim of producing reliable hazard maps.

The rainfall from the ground surface is a key factor for slope instability, as well as the pre-event degree of saturation of the soil and the interaction between the slope and the atmosphere. Many debris flows begin as shallow landslides induced by direct infiltration of intense rainfall, and runoff into cover materials.

Similarly, the pore water pressure plays a role during the propagation stage. The entrainment of further material makes the propagation patterns more complex.

The dynamic approach can be numerically solved through models based on fluid mechanics. It is possible to find propagation models applicable to hyper-concentrated flows, which mainly differ in their adopted rheological schemes. Thus, the choice of rheological parameters is a fundamental aspect of the numerical simulation.

To improve knowledge of the rheological properties of debris flow, comparisons between numerical predictions, innovative laboratory tests, and results deduced by case histories or physical model simulations can be used.

In this context, the present study underlines the possibility of acquiring knowledge of the hydro-mechanical parameters, which can be used for numerical simulations in areas with a similar soil composition. Finally, matric suction, which reflects soil infiltration, can be considered very important for the analysis of rainfall-induced landslides.

Author Contributions: Conceptualization, F.C.; Funding acquisition, F.C.; Investigation, A.D.V.; Methodology, V.L.; Supervision, V.L.; Validation, V.L.; Writing—original draft, F.C. and V.L.; Writing—review & editing, V.L. All authors have read and agreed to the published version of the manuscript.

Funding: This research was funded by the MUR (Ministry of University and Research (Italy)) through the project entitled "TEMI MIRATI".

Data Availability Statement: No applicable.

Acknowledgments: The authors acknowledge the financial support received from the MUR (Ministry of University and Research (Italy)) through the project entitled "TEMI MIRATI": Critical infrastructure safety due to fast moving landslide risk (Project code: ARS01_00158/CUP B76C18001140005), financed with the PNR 2015–2020 (National Research Program). The authors authorize the MUR to reproduce and distribute reprints for governmental purposes, notwithstanding any copyright notations thereon. Any opinions, findings, and conclusions or recommendations expressed in this material are those of the authors, and do not necessarily reflect the views of the MUR.

Conflicts of Interest: The authors declare no conflict of interest.

References

1. Blatz, J.; Graham, J. A system for controlled suction in triaxial tests. *Géotechnique* **2000**, *50*, 465–469. [CrossRef]
2. Savage, S.B.; Hutter, K. The motion of a finite mass of granular material down a rough incline. *J. Fluid Mech.* **1989**, *199*, 177–215. [CrossRef]

3. Chen, H.; Lee, C.F. Numerical simulation of debris flows. *Can. Geotech. J.* **2000**, *37*, 146–160. [CrossRef]
4. Denlinger, R.P.; Iverson, R.M. Granular avalanches across irregular three-dimensional terrain: 1. Theory and computation. *J. Geophys. Res.* **2004**. [CrossRef]
5. McDougall, S.; Hungr, O. A model for the analysis of rapid landslide motion across three- dimensional terrain. *Can. Geotech. J.* **2004**, *41*, 1084–1097. [CrossRef]
6. Quecedo, M.; Pastor, M.; Herreros, M.I.; Merodo, J.F. Numerical modelling of the propagation of fast landslides using the finite element method. *Int. J. Numer. Methods Eng.* **2004**, *59*, 755–794. [CrossRef]
7. Pastor, M.; Haddad, B.; Sorbino, G.; Cuomo, S.; Drempetic, V. A depth integrated coupled SPH model for flow-like landslides and related phenomena. *Int. J. Numer. Anal. Methods Geomech.* **2009**, *33*, 143–172. [CrossRef]
8. Pirulli, M.; Pastor, M. Numerical study on entrainment of bed material into rapid landslides. *Géotechnique* **2012**, *62*, 959–972. [CrossRef]
9. Montrasio, L.; Schilirò, L. Inferences on modeling rainfall-induced shallow landslides from experimental observations on stratified soils. *Ital. J. Eng. Geol. Environ.* **2018**, *2*, 77–85. [CrossRef]
10. Herschel, W.H.; Bulkley, R. Konsistenzmessungen von Gummi-Benzollösungen. *Kolloid Z.* **1926**, *39*, 291–300. [CrossRef]
11. Bagnold, R.A. Experiments on a gravity-free dispersion of large solid spheres in a Newtonian fluid under shear. *Proc. R. Soc. Lond. A* **1954**, *225*, 49–63.
12. Takahashi, T. Mechanical characteristics of debris flow. *J. Hydraul. Div.* **1978**, *104*, 1153–1169.
13. Iverson, R.M.; Reid, M.E.; LaHusen, R.G. Debris Flow mobilization from landslides. *Ann. Rev. Earth Planet. Sci.* **1997**, *25*, 85–138. [CrossRef]
14. Castelli, F.; Lentini, V. Landsliding events triggered by rainfalls in the Enna area (South Italy). In Proceedings of the Second World Landslide Forum, Roma, Italy, 3–7 October 2011; Springer: Berlin/Heidelberg, Germany, 2013; Volume 2, pp. 39–47. ISBN 978-3-642-31444-5. [CrossRef]
15. Di Grande, A.; Giandinoto, V. Plio-Pleistocene sedimentary facies and their evolution in centre-south-eastern Sicily: A working hypothesis. *Eur. Geosci. Union* **2002**, *1*, 211–221. [CrossRef]
16. Abu-Zeid, N.; Santarato, G. On the correspondence between resistivity and texture of loose sediments saturated with salt water. *Near Surf. Geophys.* **2004**, 144–149. [CrossRef]
17. Fredlund, D.G.; Rahardjo, H. *Soil Mechanics for Unsaturated Soils*; John Wiley and Sons: Hoboken, NJ, USA, 1993.
18. Vanapalli, S.K.; Fredlund, D.G.; Pufahl, D.E.; Clifton, A.W. Model for the prediction of shear strength with respect to soil suction. *Can. Geotech. J.* **1996**, *33*, 379–392. [CrossRef]
19. Koorevaar, P.; Menelik, G.; Dirksen, C. *Elements of Soil Physics*; Elsevier: Amsterdam, The Netherlands, 1983; Volume 13.
20. von Terzaghi, K. The shearing resistance of saturated soils and the angle between the planes of shear. *First Int. Conf. Soil Mech.* **1936**, *1*, 54–56.
21. Fredlund, D.G.; Morgenstern, N.R.; Widger, R.A. The shear strength of unsaturated soils. *Can. Geotech. J.* **1978**, *15*, 313–321. [CrossRef]
22. Escario, V.; Juca, J.; Coppe, M.S. Strength and deformation of partly saturated soils. In Proceedings of the 12th International Conference on Soil Mechanics and Foundation Engineering, Rio de Janeiro, Brazil, 13–18 August 1989; Volume 3, pp. 43–46.
23. Thu, T.M.; Rahardjo, H.; Leong, E.C. Shear strength and pore water pressure characteristics during constant water content triaxial tests. *J. Geotech. Geoenviron. Eng.* **2006**, *132*, 411–419. [CrossRef]
24. Hilf, J.W. An Investigation of Pore-Water Pressure in Compacted Cohesive Soils. Ph.D. Dissertation, U.S. Department of the Interior, Bureau of Reclamation, Design and Construction Division, Denver, CO, USA, 1956.
25. Wheeler, S.J.; Sivakumar, V. *Development and Application of a Critical State Model for Unsaturated Soil. Predictive Soil Mechanics*; Houlsby, G.T., Schofield, A.N., Eds.; Thomas Telford: London, UK, 1993; pp. 709–728.
26. O'Brien, J.S. Physical Processes, Rheology and Modeling of Mudflows. Ph.D. Thesis, Colorado State University, Fort Collins, CO, USA, 1986.
27. O'Brien, J.S.; Julien, P.Y. Laboratory analysis of mudflow properties. *J. Hydrol. Eng.* **1988**, *114*, 877–887. [CrossRef]
28. Bertolo, P.; Wieczorek, G.F. Calibration of numerical models for small debris flows in Yosemite Valley, California, USA. *Nat. Hazards Earth Syst. Sci.* **2005**, *5*, 993–1001. [CrossRef]
29. Boniello, M.A.; Calligaris, C.; Lapasin, R.; Zini, L. Rheological investigation and simulation of a debris-flow event in the Fella watershed. *Nat. Hazards Earth Syst. Sci.* **2010**, *10*, 989–997. [CrossRef]
30. Wu, Y.H.; Liu, K.F.; Chen, Y.C. Comparison between FLO-2D and Debris-2D on the application of assessment of granular debris flow hazards with case study. *J. Mt. Sci.* **2013**, *10*, 293–304. [CrossRef]
31. Stancanelli, L.M.; Foti, E. A comparative assessment of two different debris flow propagation approaches. Blind simulations on a real debris flow event. *Nat. Hazards Earth Syst. Sci.* **2015**, *15*, 735–746. [CrossRef]
32. Armento, M.C.; Genevois, R.; Tecca, P.R. Comparison of numerical models of two debris flows in the Cortina d'Ampezzo area, Dolomites, Italy. *Landslides* **2008**, *5*, 143–150. [CrossRef]
33. Nocentini, M.; Tofani, V.; Gigli, G.; Fidolini, F.; Casagli, N. Modeling debris flows in volcanic terrains for hazard mapping: The case study of Ischia Island (Italy). *Landslides* **2014**. [CrossRef]
34. O'Brien, J.D. *FLO-2D User's Manual*; Version 2007, Proc. of 06; FLO Engineering: Nutrioso, AZ, USA, 2007.

35. Castelli, F.; Castellano, E.; Contino, F.; Lentini, V. A web-based GIS system for landslide risk zonation: The case of Enna area (Italy). In Proceedings of the 12th International Symposium on Landslides—ISL2016, Napoli, Italy, 12–19 June 2016; Taylor & Francis: Abingdon, UK, 2016. ISBN 978-113802988-0. [CrossRef]

36. Castelli, F.; Freni, G.; Lentini, V.; Fichera, A. Modelling of a debris flow event in the Enna area for hazard assessment. In Proceedings of the I International Conference on the Material Point Method, MPM 2017, Delft, The Netherlands, 10–13 January 2017. [CrossRef]

37. Lentini, V.; Distefano, G.; Castelli, F. Consequence analyses induced by landslides along transport infrastructures in the Enna area (South Italy). *Bull. Eng. Geol. Environ.* **2018**. [CrossRef]

38. Lentini, V.; Castelli, F.; Distefano, G. Management of Transport Infrastructures: A Procedure to Assess the Landslide Risk. In *Geotechnical Research for Land Protection and Development*; Proc. CNRIG 2019; Springer: Berlin/Heidelberg, Germany, 2020. [CrossRef]

39. Rianna, G.; Pagano, L.; Urciuoli, G. Rainfall patterns triggering shallow flow-slides in pyroclastic soils. *Eng. Geol.* **2014**, *174*, 22–35. [CrossRef]

40. Iverson, R.M. Landslide triggering by rain infiltration. *Water Resour. Res.* **2000**, *36*, 1897–1910. [CrossRef]

41. Casadei, M.; Dietrich, W.E.; Miller, N.L. Testing a model for predicting the timing and location of shallow landslide initiation in soil-mantled landscapes. *Earth Surf. Process. Land.* **2003**, *28*, 925–950. [CrossRef]

42. Napolitano, E.; Fusco, F.; Baum, R.L.; Godt, J.W.; De Vita, P. Effect of antecedent-hydrological conditions on rainfall triggering of debris flows in ash-fall pyroclastic mantled slopes of Campania (southern Italy). *Landslides* **2016**, *13*, 967–983. [CrossRef]

43. Gitirana, G.F.N., Jr. Weather-Related Geo-Hazard Assessment Model for Railway Embankment Stability. Ph.D. Thesis, University of Saskatchewan, Saskatoon, SK, Canada, 2005; 411p.

44. Briggs, K.M.; Smethurst, J.A.; Powrie, W.; O'Brien, A.S. The influence of tree root water uptake on the long term hydrology of a clay fill railway embankment. *Transp. Geotech.* **2016**, *9*, 31–48. [CrossRef]

45. Reder, A.; Rianna, G.; Pagano, L. Some aspects of water and energy budget of a pyroclastic cover. *Environ. Geotech.* **2019**, *6*, 406–419. [CrossRef]

46. Armanini, A.; Fraccarollo, L.; Rosatti, G. Two-dimensional simulation of debris flows in erodible channels. *Comput. Geosci.* **2009**, *35*, 993–1006. [CrossRef]

47. Armanini, A.; Capart, H.; Fraccarollo, L.; Larcher, M. Rheological stratification in experimental free-surface flows of granular-liquid mixtures. *J. Fluid Mech.* **2005**, *532*, 269–319. [CrossRef]

geosciences

Article

The Hydromechanical Interplay in the Simplified Three-Dimensional Limit Equilibrium Analyses of Unsaturated Slope Stability

Panagiotis Sitarenios [1],* and Francesca Casini [2]

[1]   School of Energy, Construction and Environment, Coventry University, Coventry CV1 5FB, UK
[2]   Dipartimento di Ingegneria Civile e Ingegneria Informatica (DICII), Università degli Studi di Roma "Tor Vergata", 00133 Roma, Italy; francesca.casini@uniroma2.it
*   Correspondence: panagiotis.sitarenios@coventry.ac.uk

Abstract: This paper presents a three-dimensional slope stability limit equilibrium solution for translational planar failure modes. The proposed solution uses Bishop's average skeleton stress combined with the Mohr–Coulomb failure criterion to describe soil strength evolution under unsaturated conditions while its formulation ensures a natural and smooth transition from the unsaturated to the saturated regime and vice versa. The proposed analytical solution is evaluated by comparing its predictions with the results of the Ruedlingen slope failure experiment. The comparison suggests that, despite its relative simplicity, the analytical solution can capture the experimentally observed behaviour well and highlights the importance of considering lateral resistance together with a realistic interplay between mechanical parameters (cohesion) and hydraulic (pore water pressure) conditions.

Keywords: unsaturated slope; Ruedlingen field experiment; lateral resistance; limit equilibrium solution

Citation: Sitarenios, P.; Casini, F. The Hydromechanical Interplay in the Simplified Three-Dimensional Limit Equilibrium Analyses of Unsaturated Slope Stability. *Geosciences* **2021**, *11*, 73. https://doi.org/10.3390/geosciences11020073

Academic Editors: Jesus Martinez-Frias, Roberto Vassallo, Piernicola Lollino, Luca Comegna and Roberto Valentino
Received: 28 December 2020
Accepted: 25 January 2021
Published: 8 February 2021

Publisher's Note: MDPI stays neutral with regard to jurisdictional claims in published maps and institutional affiliations.

1. Introduction

It is well known that partial saturation can play a key role in stabilizing both natural and artificial slopes. The interparticle forces due to the formation of water menisci in the macrostructure and/or intermolecular forces due to absorptive and osmotic phenomena increase the shear strength of geomaterials, improving stability conditions. Loss of this beneficial contribution can significantly hinder stability conditions and trigger slope failures. A characteristic example is rainfall induced landslides which comprise one of the most common geotechnical failures [1–4].

Geotechnical failures occurring due to changes in hydraulic conditions highlight the significance that unsaturated soil mechanics can have in engineering problems. Unfortunately, despite its significant importance in many geotechnical problems, unsaturated soil mechanics still struggles today to find a place in everyday geotechnical practice. The reasons for this are usually related to the relative complexity of the material laws involved (i.e., water retention behaviour), the increased complexity of the required laboratory tests (i.e., suction controlled shear strength determination) and the lack of simple calculation tools accessible to everyday practitioners with fundamental knowledge of soil mechanics.

In this respect, the present paper presents and evaluates a simple slope stability limit equilibrium solution for translational failure modes. The proposed solution is initially obtained under two dimensions (2D) and it is later extended to three dimensions (3D) by additionally examining the potential development of shear strength along the sides of a sliding block with a finite lateral dimension. Similar solutions for unsaturated soil conditions exist in the international literature [1,5–8], where some refer to much simpler geometries (i.e., [6]) and others to more complex geometries requiring the use of the method of slices (i.e., [8]). Those closer to the mechanism adopted in this study usually involve assumptions which prevent them from remaining valid under saturated conditions, especially when positive pore water pressures appear. The most common reasons are either the use of stress parameters which cannot recall Terzaghi's effective stress upon saturation (i.e., [7]) or assumptions with respect to the lateral in situ stress and its evolution with partial saturation, which are incompatible with classical soil mechanics principles (i.e., [1,5]).

The proposed stability solution is a natural extension of classical soil mechanics which is key in facilitating the adoption of such solutions in engineering practices [4]. To evaluate the proposed solution, this is applied to the analyses of a well instrumented field experiment where a slope failure was triggered by means of artificial rainfall—the Ruedlingen experiment [1,9]. This proves that, despite its simplicity, the proposed mechanism can very satisfactorily represent the experimentally observed behaviour both qualitatively and quantitatively. The presented discussion highlights the importance of considering the lateral development of the failure mechanism (3D) and of the natural transition to the saturated domain in capturing the behaviour.

2. A Slope Stability Limit Equilibrium Solution for Translational Modes in Unsaturated Soils

This section discusses the limit equilibrium stability calculation of a simple translational slope stability mechanism. Firstly, the assumed geometry is introduced and simple actions such as the weight of the sliding block are calculated. Then, the shear resistance along the failure plane is calculated considering the effect that partial saturation has on soil shear strength. Finally, the Factor of Safety (FoS) against failure is formulated by additionally considering the lateral resistance developing at the sides of a 3D failing soil block.

2.1. The Assumed Failure Mechanism

Figure 1 shows the assumed failure mechanism. A soil layer of thickness z rests on top of another material—the latter was assumed to be a priory of infinite strength. Both the soil surface and the interface between the two materials is described by two parallel planes with an inclination θ with the horizontal axis. The interface between the two materials acts as a potential failure surface (plane) and a simple translational failure mechanism develops where the upper layer tends to slide along the interface.

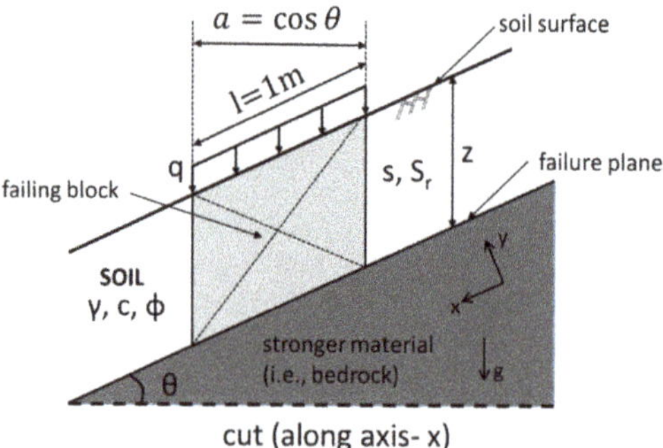

Figure 1. The analysed translational failure mechanism along a planar failure surface with slope θ.

To examine such a typical translational failure mechanism, we can isolate a block of soil with a length of $l = 1$ m along the x axis which is the one parallel to the failure plane as shown in Figure 1. To obtain an initial solution for a two-dimensional (2D) failure mechanism, an analyses per running meter was performed. The soil was assumed to be homogeneous and isotropic, while suction (s) and degree of saturation (S_r) were both assumed to be constant along the entire soil layer with no variation of depth in favour of simplicity. Moreover, for the sake of generality of the produced solution, a vertical surcharge equal to q was assumed on the surface of the soil.

The unit weight of the soil varies with degree of saturation (S_r) following Equation (1) as:

$$\gamma = \frac{G_S + e \cdot S_r}{1 + e} \cdot \gamma_w \,,$$ (1)

where G_s is the specific gravity and e the void ratio of the soil, while γ_w is the unit weigh of the water usually assumed to be equal to γ_w = 9.81 kN/m^3. Having defined the unit weight of the soil, the weight of the soil block (W) can be calculated and analysed as two components, W_x and W_y, along the x and y axes, respectively, as shown in Figure 2a:

$$W = \gamma \cdot z \cdot \cos\theta,$$ (2)

$$W_x = W \cdot \sin\theta = \gamma \cdot z \cdot \cos\theta \cdot \sin\theta,$$ (3)

$$W_y = W \cdot \cos\theta = \gamma \cdot z \cdot \cos^2\theta,$$ (4)

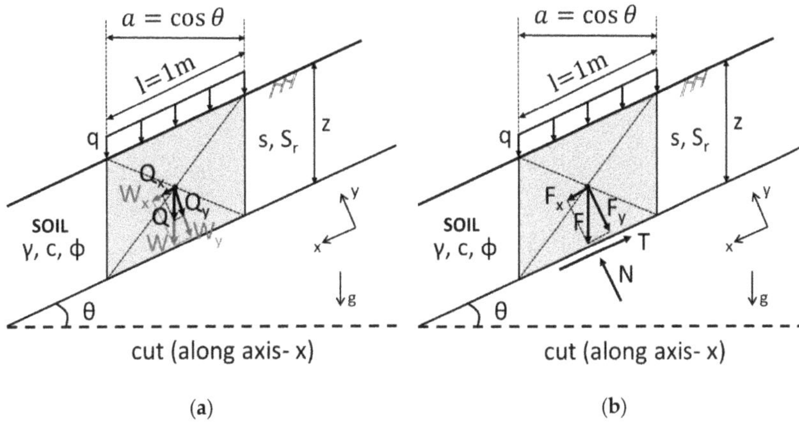

Figure 2. (**a**) Analyses of the weight of the block and of the applied surcharge; (**b**) equilibrium of the block along x axis (parallel to the failure plane) and y axis (perpendicular to the failure plane).

To further consider the effect of the uniform surcharge, we calculated the corresponding force (Q) and its two components, Q_x and Q_y, along the x and y axes, respectively, as shown in Figure 2a:

$$Q = q \cdot \cos\theta,$$ (5)

$$Q_x = Q \cdot \sin\theta = q \cdot \cos\theta \cdot \sin\theta,$$ (6)

$$Q_y = Q \cdot \cos\theta = q \cdot \cos^2\theta.$$ (7)

Finally, as portrayed in Figure 2b, the sum of the forces due to: (a) the weight of the soil block and (b) the surcharge is considered; initially on the x axis where F_x plays the role of the destabilizing action trying to slide the considered block along the failure surface,

$$F_x = Q_x + W_x = q \cdot \cos\theta \cdot \sin\theta + \gamma \cdot z \cdot \cos\theta \cdot \sin\theta = (q + \gamma z)\,\cos\theta \cdot \sin\theta,$$ (8)

and along the y-axis where F_y is equal to the reaction force from the bottom layer:

$$N = F_y = Q_y + W_y = q \cdot \cos^2\theta + \gamma \cdot z \cdot \cos^2\theta = (q + \gamma z)\,\cos^2\theta.$$ (9)

2.2. Shear Resistance under Unsaturated Conditions

The shear resistance at the interface is assumed to be entirely controlled by the shear strength parameters of the soil block and is described by the Mohr–Coulomb failure envelope. The effect of partial saturation on shear strength is considered by using Bishop's (average) skeleton stress [10]. Bishop's average skeleton stress (σ') is calculated according to Equation (10) as the sum of the total stress (σ) plus suction ($s = -u_w$, assumed air pressure: $u_a = 0$); the latter is scaled for parameter χ.

$$\sigma' = \sigma + s \cdot \chi, \tag{10}$$

Note that although the prime symbol is used for Bishop's average skeleton stress, this should not be considered an effective stress variable; in fact, a second stress variable must be defined to adequately describe the behaviour under unsaturated conditions—e.g., suction [11,12]. If χ is properly selected to correspond to $\chi = 1$ upon full saturation ($S_r = 1.0$) and to $\chi = 0$ for dry states ($S_r = 0.0$), then Bishop's skeleton stress recalls Terzaghi's effective stress upon saturation which is a desirable feature for the unified description of saturated and unsaturated material states.

The Bishop's skeleton stress combined with common failure criteria for soils proves adequate when the interest focuses exclusively on the evolution of shear strength with partial saturation. This is supported by an ensemble of different studies [13–16], all of them suggesting that χ must be a function of S_r which ensures that $\chi = 1$ for $S_r = 1.0$ and $\chi = 0$ for $S_r = 0.0$, as previously discussed.

To this end, for the examined sliding block failure mechanism the shear resistance along the interface is described by the Mohr–Coulomb failure criterion combined with Bishop's skeleton stress:

$$\tau = c + \{\sigma + s \cdot \chi\} \cdot tan\phi \tag{11}$$

where τ is the shear strength of the material described by the well known shear strength parameter cohesion (c) and angle of internal friction (ϕ). The failure criterion was then applied at the failure surface of the examined stability mechanism to calculate the shear resistance force (T) (see also Figure 2b), which can be calculated as (also see Appendix A):

$$T = c + (q + \gamma z) \, cos^2\theta \cdot tan\phi + s \cdot \chi \cdot tan\phi \tag{12}$$

Finally, the safety factor against sliding of the block can be evaluated as the ratio of the forces resisting potential sliding and those trying to destabilize the examined block in the direction of the x axis. After some algebraic calculations for the 2D failure mechanism, the Factor of Safety (FoS) can be calculated according to:

$$FoS = \frac{T}{F_x} = \frac{\frac{c}{cos^2\theta} + \frac{s \cdot \chi \cdot tan\phi}{cos^2\theta} + (q + \gamma z) \, tan\phi}{(q + \gamma z) tan\theta} \tag{13}$$

2.3. Extension to 3D-Lateral Shear Resistance

The simple 2D translational slope failure mechanism presented hereinbefore and quantified through Equation (13) can be used to evaluate the stability of planar slope instabilities where failure extends enough to reasonably assume that stability can be evaluated by performing a "per running meter analyses" where a slice of the falling mass with a thickness of 1m at the centre of the failure area is isolated and studied. However, in many cases, slope failures have rather finite dimensions where the aforementioned assumption becomes problematic.

In such cases, the main shortcoming of the 2D mechanism is that it neglects the contribution of the shear resistance developing at the sides of the failing mass. Neglecting their contribution proves conservative and usually this is the justification accompanying the assumption of simple 2D slope stability methods in the design of slopes and cuts. However, this is not true when it comes to the back analyses of slope failures where neglecting the actual dimensions of the failure mass and the contribution of the lateral resistance can significantly hinder the outcome of the analyses (i.e., overestimation of soil shear strength parameters and/or suction contribution).

To this end, this section extends the discussed translational failure mechanism in the 3D domain by additionally considering the contribution of lateral resistance. For this, the geometry presented in Figure 3 is considered. The failing block has a width (b) which limits its lateral development, while the depth of the failure mass is (z) and the length is $l = 1$ m, which as a horizontal distance translates to $a = \cos\theta$, similar to the 2D mechanism.

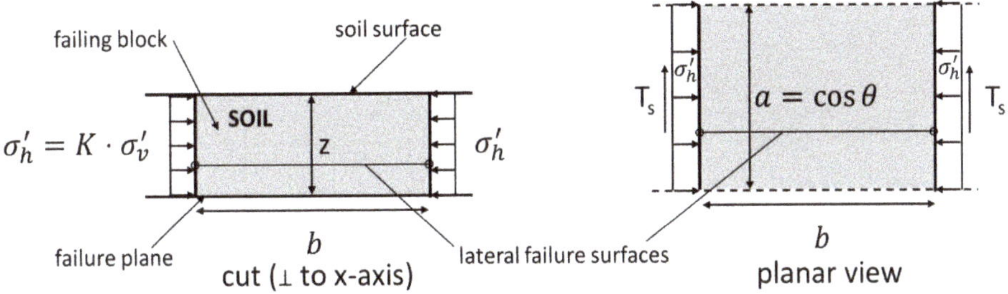

Figure 3. The 3D failure mechanism in cut and planar view.

To calculate the shear resistance T_s developing along each of the two lateral sides of the block, the Mohr–Coulomb failure criterion combined with Bishop's average skeleton stress is used again. The horizontal stress at the two lateral interfaces is assumed to be equal to $\sigma'_h = K \cdot \sigma'_v$ where K is the coefficient of lateral earth pressure. Applying the lateral earth pressure coefficient to Bishops' stress as opposed to total stress that other solutions assume (i.e., [1]) makes the derived solution applicable under positive pore water pressure regimes as well, while recent experimental evidence on soil anisotropy and partial saturation [17] also support such an assumption. The vertical Bishop's skeleton stress is calculated in the middle of the depth (z) as:

$$\sigma'_v = \sigma_v + s \cdot \chi = q + \gamma \cdot \frac{z}{2} + s \cdot \chi \tag{14}$$

and the corresponding horizontal stress as:

$$\sigma'_h = K \cdot \sigma'_v = K \cdot q + K \cdot \gamma \cdot \frac{z}{2} + K \cdot s \cdot \chi \tag{15}$$

The stress as calculated in the middle of the layer is assumed uniform along the lateral sides of the excavation and the shear strength of the soil is then calculated as:

$$\tau_s = c_s + \sigma'_h \cdot tan\phi_s \tag{16}$$

where c_s and ϕ_s are the cohesion and frictional resistance, respectively, along the lateral sides of the failure block. They are assumed to be different from their soil counterparts in favour of generality of the solution, by allowing different shear strength parameters to be assigned to the lateral resistance. Having defined the shear strength along the two lateral sides, we can calculate the corresponding shear resistance, where after some algebraic calculations (see Appendix B) the following expression is derived:

$$T_s = c_s \cdot z \cdot cos\theta + K \cdot q \cdot z \cdot cos\theta \cdot tan\phi_s + K \cdot \gamma \cdot \frac{z^2}{2} \cdot cos\theta \cdot tan\phi_s + K \cdot s \cdot \chi \cdot z \cdot cos\theta \cdot tan\phi_s \tag{17}$$

Finally, using the additional resistance along the sides of the 3D mechanism, the safety factor against sliding is reformulated:

$$FoS = \frac{T + \frac{2T_s}{b}}{F_x} \tag{18}$$

Note that the additional resistance at the two sides ($2T_s$) is normalized over the width (b) of the block to combine it mathematically with the already determined resistance T and action F_x for the 2D mechanism. Finally, by using Equations (8), (12) and (17) to substitute for F_x, T and T_s, respectively, in Equation (18), we can calculate that the Factor of Safety (FoS) against sliding for the 3D mechanism as:

$$FoS = \frac{\frac{c}{cos^2\theta}\left(1 + \frac{2\rho_c cos\theta}{b}\right) + \frac{s\chi tan\phi}{cos^2\theta}\left(1 + \frac{2\rho_\phi Kz cos\theta}{b}\right) + q \cdot tan\phi\left(1 + \frac{2\rho_\phi Kz}{b \cdot cos\theta}\right) + \gamma z \cdot tan\phi\left(1 + \frac{\rho_\phi Kz}{b \cdot cos\theta}\right)}{(q + \gamma z) \cdot tan\theta} \tag{19}$$

In Equation (19), and in order to obtain a more elegant solution, the shear strength parameters used for the resistance along the inclined failure surface and along the sides of the block are correlated using the following factors:

$$\rho_s = \frac{c_s}{c} \tag{20}$$

$$\rho_\varphi = \frac{tan\phi_s}{tan\phi} \tag{21}$$

3. Evaluation Using an Actual Case

To evaluate the planar slope stability mechanism previously presented and to demonstrate the potential importance of lateral resistance in analysing slope instabilities, the results from the Ruedlingen full scale experiment [1,18] will be used and compared with the predictions of the proposed stability mechanism.

3.1. The Ruedlingen Full Scale Experiment

The Ruedlingen full scale field test [1,18] was performed to study the response of a steep forested slope, subjected to artificial intense rainfall within the context of the multidisciplinary research programme on the "Triggering of Rapid Mass Movements in steep terrain" (TRAMM). The experiment was carried out in northern Switzerland in a forested area near Ruedlingen village. The selected experimental site is located on the east-facing bank of the river Rhine, with an average slope angle of approximately 38°. An orthogonal area, with a length of 35 m and a width of 7.5 m, was instrumented with a wide range of devices including earth pressure cells, piezometers, tensiometers, time domain reflectometers (TDRs), and acoustic and temperature sensors, organized in clusters, as seen in Figure 4b. A landslide triggering experiment was conducted in March 2009 where the slope was subjected to artificial rainfall with an average intensity of 20 mm/h at the upper one-third (approximately) of the slope and with 7 mm/h in the remaining lower part leading to a shallow slope failure after 15 h.

The failure concentrated at the upper part of the instrumented area (see Figure 4b) and extended within a surficial layer of a medium to low plasticity (average PI ∼10%) silty sand (ML) [19]. The developed failure mechanism can be characterized as a surficial planar failure. The mobilized area had a length of ∼17 m, a width of ∼7 m while the failure surface can be practically assumed to be parallel to both the soil surface and the underlying soil–bedrock interface with its maximum depth exciding 1 m [18]. This soil layer rests on top of a fissured Molassic bedrock (sandstone and marlstone—see Figure 4a) [20,21]. Sitarenios et al. [9] numerically analysed the triggered slope instability and, by combining field measurements with numerical results, they concluded that failure happens under saturated conditions and is triggered by water exfiltration from the bedrock which results in considerable pore water pressure build-up at the soil–bedrock interface.

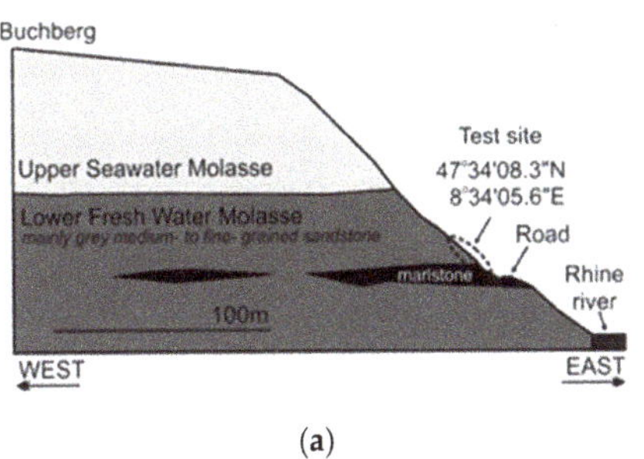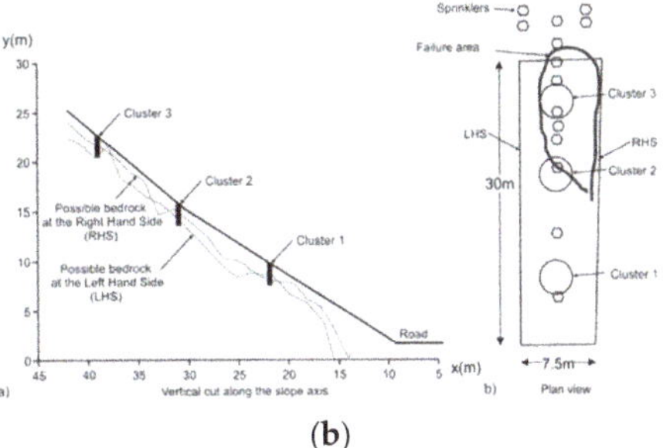

Figure 4. (**a**) Geomorphology of the greater experimental area (after [20]); (**b**) a schematic of the bedrock topography and of the instrumented area of the experiment (after [22]).

3.2. 2D Solution—Neglecting Lateral Resistance

In this section, the proposed limit equilibrium mechanism is used to evaluate the stability of the Ruedlingen slope by neglecting any contribution of lateral resistance or, in other words, by assuming an infinite failure mechanism perpendicular to the examined plane. When utilizing the mechanism, it is important to first establish the hydraulic and mechanical parameters of the Ruedlingen soil; to enable comparison of the results with the study of Sitarenios et al. [9], the same experimental data will be examined and similar assumptions will be put forward.

Starting with the hydraulic behaviour and according to what was previously discussed, calculation of the soil shear strength under unsaturated conditions necessitates knowledge of the degree of saturation as the latter is the most frequently made assumption for parameter (χ). To calculate the degree of saturation and thus parameter (χ), the Water Retention Model (WRM) of Equation (22) is used where s is the suction level, b_w is a model parameter controlling the shape of the reproduced Water Retention Curve (WRC) and $S_{r,max}$ and $S_{r,min}$ are the maximum and residual degrees of saturation, respectively. It is the a Van Genuchten type WRM [23] further including dependence of the behaviour of the void ratio ($e = n/(1-n)$) through parameter P, as described in Equation (23), where P_0 and n_0 are reference values and parameter a_w controls the rate of change of the WRC with the void ratio.

$$\chi = S_r = S_{r,res} + (S_{r,max} - S_{r,res})\left[1 + \left(\frac{s}{P}\right)^{\frac{1}{1-b_w}}\right]^{-b_w} \tag{22}$$

$$P = P_0 \cdot \exp[a_w(n_0 - n)] \tag{23}$$

It should be highlighted that the water retention behaviour is key to the accurate representation of the unsaturated shear strength. Consequently, in case of significant hysteresis in the water retention behaviour, attention must be given to the utilization of this branch of the curve (drying vs. wetting) which best represents the physically analysed problem. This study addresses a wetting problem (rainfall water infiltration) and, in this respect, the water retention behaviour is calibrated based on a set of suction controlled wetting laboratory tests [24]. The calibrated behaviour can be seen in Figure 5a and the parameters used are shown in Table 1. Figure 5b shows the WRC curve that is used in the analyses, which corresponds to an initial void ratio of $e = 0.9$ as this has been measured in the field and also used in previous studies [9]. It is accompanied by the evolution of the $s \cdot S_r$ term which in fact reflects the increase in the shear strength of the material with suction.

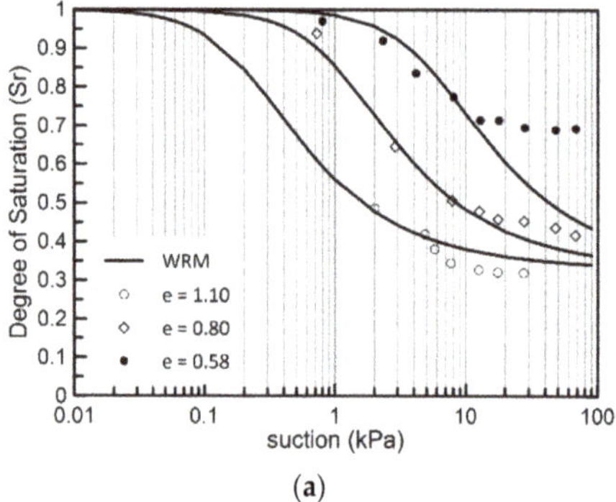

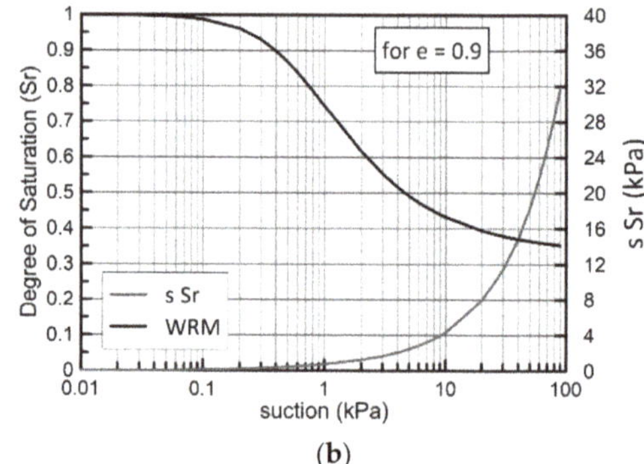

Figure 5. (**a**) The calibrated Water Retention Model (WRM) (laboratory data from [24]); (**b**) the Water Retention Curve (WRC) used in the analyses with the corresponding evolution of the $s \cdot S_r$ term.

Table 1. Ruedlingen WRM parameters used in this study.

Parameter	Value	Parameter	Value
P_0 (kPa)	0.65	α	21.0
b_w	0.4	n_0	0.47
$S_{r,max}$	1.0	$S_{r,min}$	0.33

With the WRM calibrated, Equation (11) is used to calculate the safety factor of the Ruedlingen slope under different combinations of suction and slope heights. With respect to the shear strength parameters involved in Equation (11), cohesion is set to zero ($c = 0$) and the friction angle equal to $\phi = 32$ [9,19], while the unit weight (γ) of the slope follows the evolution of saturation through Equation (1), with specific gravity equal to $G_s = 2.65$. Finally, the inclination of the failure surface is set equal to the average inclination of the slope ($\theta = 38$) [1] and no surcharge is considered on the soil surface ($q = 0$).

Figure 6 presents the results of a parametric study. Figure 6a portrays the evolution of the FoS with suction for various slope heights ranging from 0.5 to 2.75 m. The reason why the height of the slope is handled parametrically is related to the in situ variability of the thickness of the soil layer which was found, in general, to range from 0.5 to 4.5 m [18,20], while in the part of the slope where failure occurred the variation was smaller, with depths up to 2.75 m. In Figure 6b, similar results are presented with depth in the horizontal axis and the various curves represent different suction levels from zero (0) up to 10 kPa, following the in situ initial suction values as those have been reported in [18].

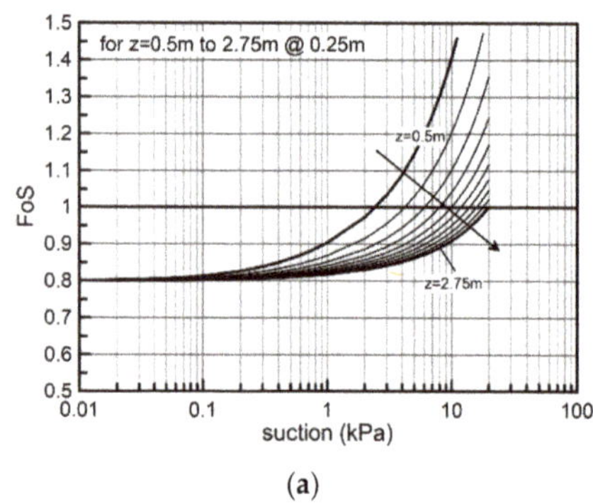

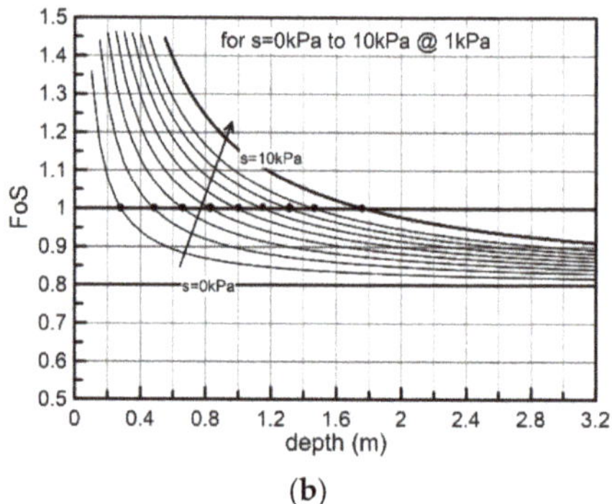

Figure 6. (**a**) Evolution of Factor of Safety (FoS) with suction for different slope heights; (**b**) evolution of FoS with slope height for different suction levels; results from a 2D mechanism (no lateral resistance).

The results of Figure 6 clearly demonstrate the dominant effect that partial saturation can have on stabilizing a planar slope. It is characteristic that when no suction is present ($s = 0$), a cohesionless slope ($c = 0$) has a safety factor which follows the $\frac{tan(\phi)}{tan(\theta)}$ ratio; the latter, for the examined slope geometryand soil shear strength is equal to $\frac{tan(32)}{tan(38)} = 0.8$. In both Figure 6a,b, it can be observed that even a small suction increase significantly alters stability conditions with various combinations of suction and slope heights exhibiting stable behaviours (FoS $\geq$ 1).

To better understand this interplay between the height of the slope and suction, Figure 7 presents stability charts correlating the two quantities, with suction represented by the horizontal axis and slope height by the vertical axis, respectively. Figure 7a suggests that there is an almost linear relation between the increase in depth and the required increase in suction for the slope to remain stable (FoS > 1), while in Figure 7b the same relation is presented using a logarithmic scale with suction. Figure 7b helps to understand that the increase in the stable slope height with suction in fact follows the trend of the increase in the shear strength with suction as the latter is controlled by the increase in the $s \cdot S_r$ term in Figure 5b.

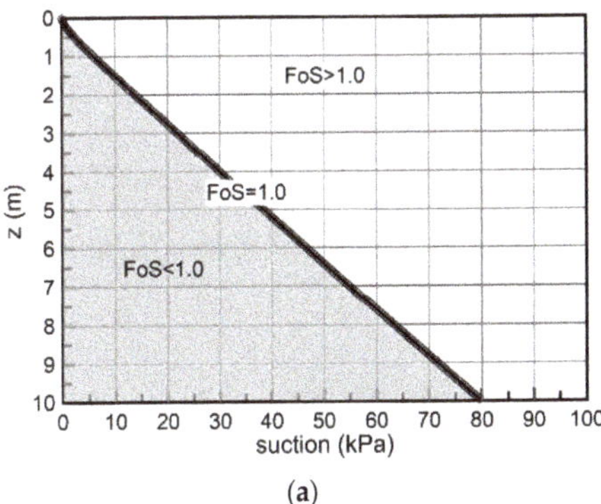

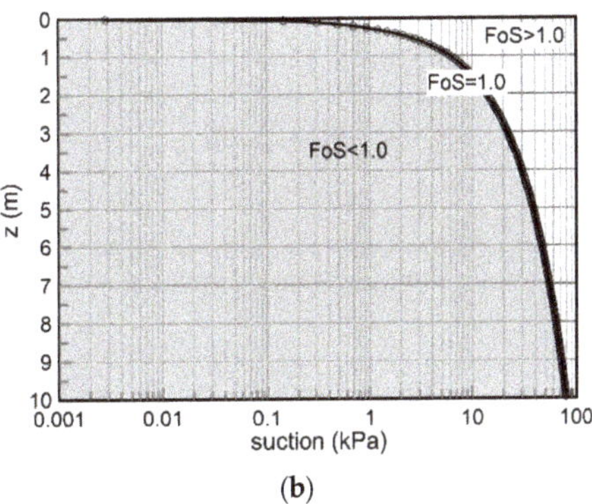

Figure 7. Stability chart; (a) stable vs. unstable suction and slope height combinations; (b) same as (a) in a semilogarithmic plot; results from a 2D mechanism (no lateral resistance).

Examining the stability chart of Figure 6 in relation to the observed behaviour during the Ruedlingen field experiment, it seems that the suggested evolution of stability with suction and slope height does not match the observed behaviour. If we consider an average slope height of 1.5 to 2 m, Figure 7 suggests that a minimum suction value of 10 to 15 kPa is required for stability. This may partially explain why the slope remains stable under the initial hydraulic regime which suggests a value of around 10 kPa, but at the same time it is obvious that even the smallest decrease in suction should cause failure. However, both in situ measurements [1] and numerical analyses [9] clearly indicate that even complete loss of suction is not enough to trigger the Ruedlingen failure. In this respect, it seems reasonable to assume that additional factors contribute to stability with respect to those examined with the present 2D failure mechanism.

3.3. 3D Solution—Including Lateral Resistance

Lateral resistance can be one of the factors contributing to improved stability conditions with respect to those of Figure 7. To consider the effect of lateral resistance in the examined problem, the analysis of the previous section is repeated by additionally considering the lateral resistance at the two sides of a failing block with a width equal to b = 7.5 m. For the side resistance, the same friction angle as for the soil–bedrock interface is assumed (ρ_ϕ=1.0), while for the lateral earth pressure coefficient an arbitrary value of K = 0.5 is chosen, close to the value suggested by Jaky's formula for $\phi = 32 \Rightarrow K = 1 - \sin 32 = 0.47$.

Figure 8 presents results similar to those of Figure 6; however, Figure 8 includes the additional contribution of lateral resistance. The results clearly demonstrate the beneficial effect of lateral resistance in increasing FoS under given combinations of slope height and suction. It is interesting to observe that Figure 8a seems to suggest that the effect of suction in improving stability conditions becomes less profound as the slope height increases. Moreover, Figure 8b indicates that under relatively small slope height values (i.e., less than 0.6 m), the influence of suction is quite significant, while as the slope height increases its effect becomes less profound, reaching a minimum for a given depth and then slightly increasing again.

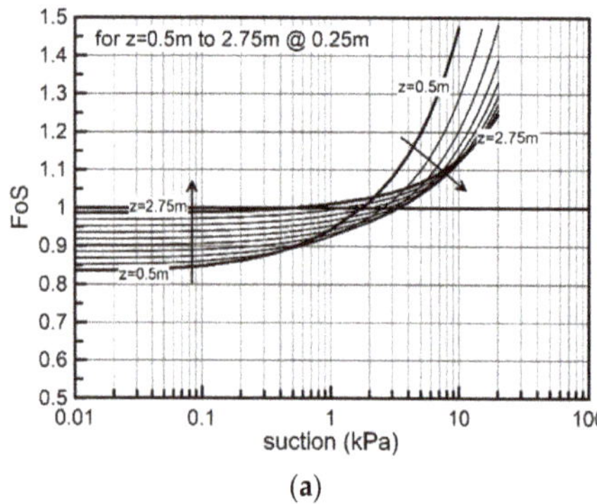

(**a**)

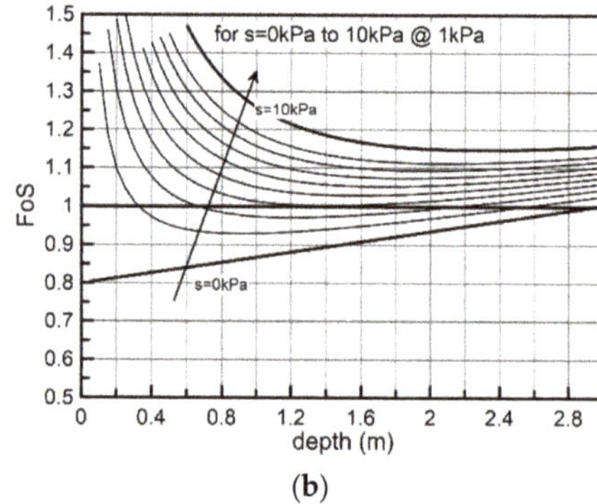

(**b**)

Figure 8. (**a**) Evolution of FoS with suction for different slope heights; (**b**) evolution of FoS with slope height for different suction levels; results from a 3D mechanism (with lateral resistance).

This effect is more profound in the results shown in Figure 9, presenting a stability chart similar to Figure 7 with the addition of lateral resistance. It is observed that as the height of the slope increases, positive pore water pressure values are needed to cause instability. Qualitatively, this trend in behaviour is much better aligned with the behaviour of the Ruedlingen slope, where failure happened following considerable water pressure build-up which reduced the available shear strength at the soil–bedrock interface. Moreover, both postfailure in situ observations and numerical results suggest that the depth of the failure surface ranged from 1.0 to 2.0 m. This is in good agreement with Figure 9, where within the same range of z values, the stability chart exhibits a peak coinciding with the most unfavourable stability conditions for the given assumptions.

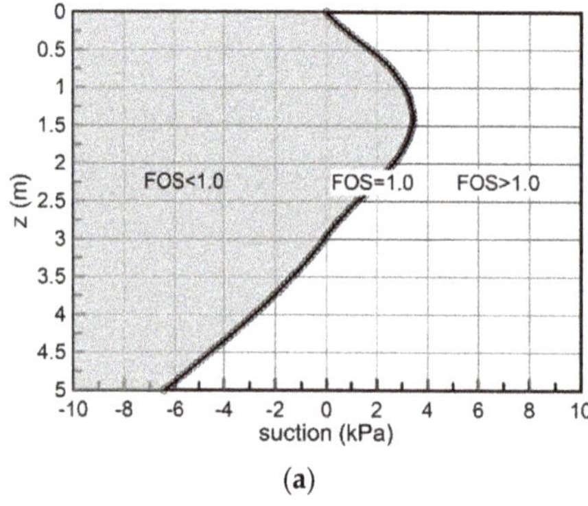

(**a**)

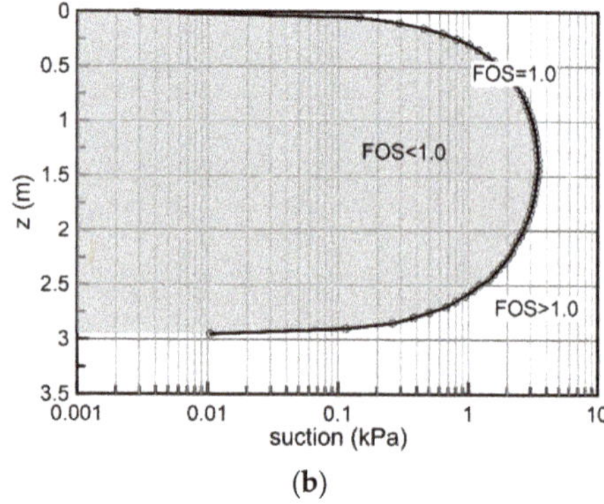

(**b**)

Figure 9. Stability chart; (**a**) stable vs. unstable suction and slope height combinations; (**b**) as (**a**) in a semilogarithmic plot; results from a 3D mechanism (with lateral resistance).

However, from a quantitative perspective and within the same range of slope heights, the required suction for stability suggests that the slope should have failed before reaching full saturation contrary to what was observed in situ. This discrepancy can be attributed to several factors which can result in better stability conditions than those described by the analysis used for this study. Amongst others, three of the most likely potential candidates for the different behaviours can be (a) the presence of some cohesion either due to fines or due to the contribution of vegetation and roots [25]; (b) a less extended mechanism in the third direction (smaller b) and (c) a different WRC. However, it should be recognized that the WRC can only affect the behaviour in the unsaturated regime and still cannot explain why a higher positive pore water pressure may be needed to trigger the instability.

3.4. 3D Solution—The Effect of Cohesion and 3D Mechanism Development

The previous section demonstrated how the inclusion of the lateral resistance in the analyses can justify an improvement of the stability conditions for the analysed Ruedlingen field experiment. However, it was not feasible to quantitatively approach the behaviour satisfactorily. As mentioned before, either the presence of cohesion or a narrower than assumed mechanism can serve as the most probable explanations and this hypothesis is put to test in this section.

To examine the effect of different assumptions regarding the aforementioned factors, a parametric analysis is conducted to check the sensitivity of the stability charts on the presence of cohesion (c) and on the width (b) of the 3D mechanism. Figure 10a presents how the stability curve differentiating between stable and unstable conditions changes with an increase in cohesion. Five different cohesion values, namely $c = 0, 1, 2, 3, 4$ and 5 kPa (from the blue to the red curve), are examined and the same cohesion values also control the lateral resistance by assuming that $\rho_c = 1.0$ (Equation (20)). The plotted charts indicate a significant contribution of cohesion to the stability conditions. Even the smallest nonzero value of $c = 1$ kPa is enough to justify a stable slope under the presence of the smallest suction and irrespective of its height. This is in accordance with the experimentally observed behaviour, while the presence of roots can justify even higher values of cohesion [25] which, if adopted (i.e., $c = 5$ kPa), can explain why significant pore pressures up to almost 10 kPa were needed in the soil–bedrock interface for failure to occur in the Ruedlingen experiment.

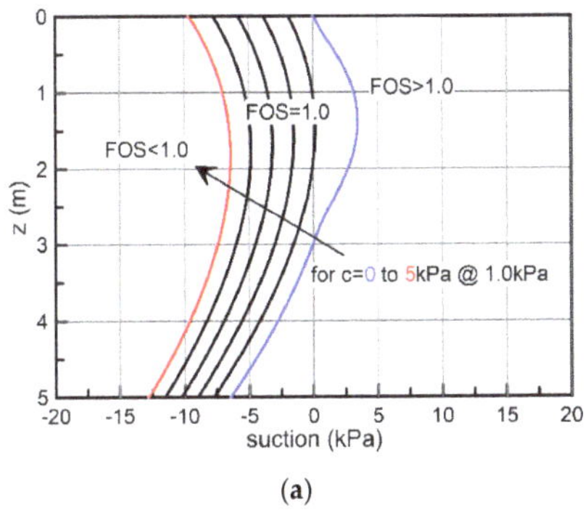

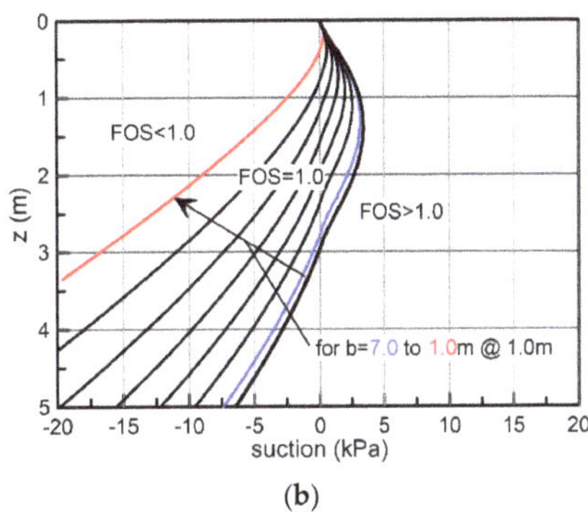

Figure 10. (a) Effect of cohesion on the stability chart; (b) effect of the 3D mechanism width on the stability chart; results from a 3D mechanism (with lateral resistance).

Figure 10b offers a similar discussion by explaining the effect of the width of the assumed mechanism. The main assumption put forward in the analyses presented hitherto is that the width of the mechanism is $b = 7.5$ m, coinciding with the width of the experimental area (dashed curve on Figure 9b). However, the in situ postfailure observations suggest that the failing mass had a smaller average width (see also Figure 4b), and the results of Figure 10b indicate that such a reduction in the lateral development of the failure mechanism is quite beneficial for stability. In more detail, Figure 10b presents the evolution of the stability curves with the width of the mechanism, namely from $b = 7$ m down to $b = 1$ m (from the blue to the red curve). It is interesting that the effect of lateral development was qualitatively different compared to that of cohesion, with the slope height playing a key role. It is observed that as the slope becomes higher and the mechanism narrower, the effect of the lateral resistance on improving stability becomes higher. This is a clear reflection of the fact that the ratio of the area of the lateral side over the area of the soil–bedrock interface becomes higher, increasing the contribution of the side resistance and limiting the contribution of the failure plane resistance, thus progressively making the lateral resistance more and more dominant.

Regarding the in situ behaviour, it is clear that a narrower mechanism cannot explain the in situ behaviour alone; as for the slope height relevant to the Ruedlingen soil ($z = 1$ to 2 m) and the range of the lateral development of the mechanism ($b = 5$ to 7 m), the slope is still unstable, even though there is a noticeable improvement in stability conditions with respect to the refence scenario ($b = 7.5$ m). In this respect, the best candidate to justify the experimental behaviour is cohesion. From the results of Figure 10, it seems that reasonable assumptions regarding a small presence of cohesion in the order of $c = 3$ to 4 kPa, combined with a relatively smaller average mechanism $b = 5$ to 7 m, can explain very satisfactory the experimentally observed behaviour during the Ruedlingen experiment, both qualitatively and quantitively.

To test this assumption, four combinations of cohesion and slope width within the aforementioned range of values are finally analysed Figure 11b. In the same graph, the grey shaded area indicates the range of values of slope height and "suction" (negative suction values correspond to positive pore water pressure) most representative of the in situ failure conditions. In more detail, Figure 11a presents the failure mechanism (after [1]) where the average depth of the failure surface was more than 1 m. It also contains a plot of the evolution of the in situ and of the numerically determined pore water pressure at the failure area (after [9]) indicating that an average pore water pressure in the order of 5 kPa was present in the slope mass at failure. The very good match of the stability curves with the shaded area is a clear indication that the relatively simple 3D translational failure mechanism used in this study can provide a very good insight into the Ruedlingen failure.

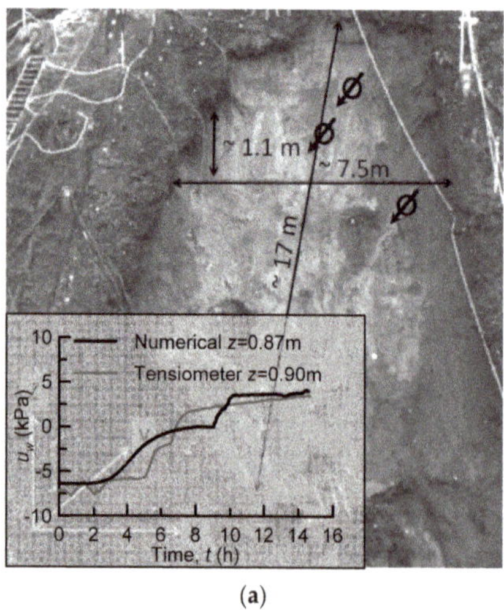

(**a**)

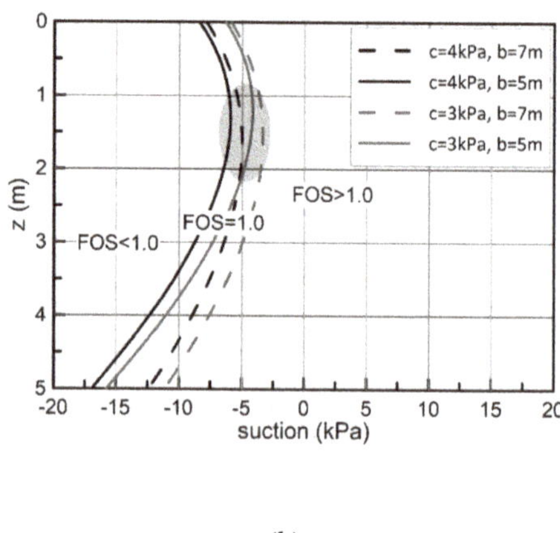

(**b**)

Figure 11. (**a**) The Ruedlingen slope failure (after [1]) and the evolution of pore water pressure in the slope mass (after [9]); (**b**) stability curves for different combinations of cohesion and slope widths which can accurately capture the observed instability (grey shaded area).

4. Conclusions

In this paper, a simplified limit equilibrium slope stability solution for infinite translational failure mechanisms was introduced. The solution uses Bishop's average skeleton stress to account for the increase in the shear strength under unsaturated conditions. It was extended in 3D dimensions by considering a potential limited lateral development of the failure mechanism by accounting for the lateral shear resistance that develops across the sides of the failing mass. Contrary to other similar attempts, the proposed solution assumes that the lateral earth pressure coefficient (K) describes the ratio of the horizontal over the vertical Bishop's stress and not only the net stress ratio. By doing so, the solution remains valid under fully saturated conditions where Bishop's average skeleton stress recalls Terzaghi's effective stress.

The mechanism was evaluated using a set of actual data from a full scale field experiment—the Ruedlingen artificial rainfall induced landslide. The proposed stability solution, despite its simplicity, can capture very satisfactory the experimentally observed behaviour. Additionally, its simplicity and very limited calculation costs enable an easy evaluation of different scenarios with respect to slope geometries and soil parameters which can prove very useful in back analysis of slope failures. Such simple validated tools for the analyses of engineering problems under unsaturated conditions, formulated as a natural extension of similar classical soil mechanics solutions, are very important in facilitating a wider adoption of unsaturated mechanics in everyday geotechnical practice.

Author Contributions: Conceptualization, F.C.; Investigation, P.S.; Methodology, P.S.; Writing—original draft, P.S.; Writing—review & editing, F.C. All authors have read and agreed to the published version of the manuscript.

Funding: This research received no external funding.

Informed Consent Statement: Not applicable.

Data Availability Statement: Not applicable.

Conflicts of Interest: The authors declare no conflict of interest.

Appendix A

Starting from the Mohr–Coulomb criterion, Equation (11) is multiplied with the area of the failure surface, which is equal to $l = 1$ m, by 1 m for the distance perpendicular to the examined 2D problem (analyses per running meter. Thus, the shear resistance as a force is derived

$$T = c \cdot l + \{\sigma_n \cdot l + s \cdot \chi \cdot l\} \cdot tan\phi \text{ (for } l = 1 \text{ m) or} \tag{A1}$$

$$T = c + N \cdot tan\phi + s \cdot \chi \cdot tan\phi \tag{A2}$$

In Equation (A2), N represents the reaction force at the failure interface as this has been calculated with Equation (9) and so we can finally write:

$$T = c + (q + \gamma z) \, \cos^2 \theta \cdot tan\phi + s \cdot \chi \cdot tan\phi \tag{A3}$$

Appendix B

In Equation (16), which represents the shear strength along the side of the examined mechanism, Bishop's average skeleton stress is described through Equation (15) to obtain:

$$\tau_s = c_s + \sigma'_h \cdot tan\phi_s = c_s + \left(K \cdot q + K \cdot \gamma \cdot \frac{z}{2} + K \cdot s \cdot \chi \right) tan\phi_s \tag{A4}$$

Focusing on one side of the 3D mechanism, the shear strength above develops along the lateral side of the mechanism which is portrayed in Figure 1 by the light grey shaded area. It is a parallelogram with an area equal to $z \cdot a = z \cdot cos\theta$. Multiplying Equation (A4) with $z \cdot cos\theta$, the shear resistance along the side is:

$$T_s = c_s \cdot z \cdot cos\theta + \left(K \cdot q + K \cdot \gamma \cdot \frac{z}{2} + K \cdot s \cdot \chi \right) z \cdot cos\theta \cdot tan\phi_s \text{ or} \tag{A5}$$

$$T_s = c_s \cdot z \cdot cos\theta + K \cdot q \cdot z \cdot cos\theta \cdot tan\phi_s + K \cdot \gamma \cdot \frac{z^2}{2} \cdot cos\theta \cdot tan\phi_s + K \cdot s \cdot \chi \cdot z \cdot cos\theta \cdot tan\phi_s \tag{A6}$$

References

1. Askarinejad, A.; Casini, F.; Bischof, P.; Beck, A.; Springman, S.M. Rainfall Induced Instabilities: A Field Experiment on a Silty Sand Slope in Northern Switzerland. *RIG Ital. Geotech. J.* **2012**, *3*, 50–71.
2. Tagarelli, V.; Cotecchia, F. The Effects of Slope Initialization on the Numerical Model Predictions of the Slope-Vegetation-Atmosphere Interaction. *Geosciences* **2020**, *10*, 85. [CrossRef]
3. Cotecchia, F.; Santaloia, F.; Tagarelli, V. Towards A Geo-Hydro-Mechanical Characterization of Landslide Classes: Preliminary Results. *Appl. Sci.* **2020**, *10*, 7960. [CrossRef]
4. Tang, A.M.; Hughes, P.N.; Dijkstra, T.A.; Askarinejad, A.; Brenčič, M.; Cui, Y.J.; Diez, J.J.; Firgi, T.; Gajewska, B.; Gentile, F.; et al. Atmosphere-Vegetation-Soil Interactions in a Climate Change Context; Impact of Changing Conditions on Engineered Transport Infrastructure Slopes in Europe. *Q. J. Eng. Geol. Hydrogeol.* **2018**, *51*, 156–168. [CrossRef]
5. Casini, F.; Minder, P.; Springman, S.M. Shear Strength of an Unsaturated Silty Sand. Unsaturated Soils. In Proceedings of the Fifth International Conference on Unsaturated Soils, Barcelona, Spain, 6–8 September 2010; pp. 211–216.
6. Springman, S.M.; Jommi, C.; Teysseire, P. Instabilities on Moraine Slopes Induced by Loss of Suction: A Case History. *Géotechnique* **2003**, *53*, 3–10. [CrossRef]
7. Zhang, L.L.; Zhang, J.; Zhang, L.M.; Tang, W.H. Stability Analysis of Rainfallinduced Slope Failure: A Review. *Proc. Inst. Civ. Eng. Geotech. Eng.* **2011**, *164*, 299–316. [CrossRef]
8. Ip, S.C.Y.; Rahardjo, H.; Satyanaga, A. Three-Dimensional Slope Stability Analysis Incorporating Unsaturated Soil Properties in Singapore. *Georisk Assess. Manag. Risk Eng. Syst. Geohazards* **2020**, 1–15. [CrossRef]
9. Sitarenios, P.; Casini, F.; Askarinejad, A.; Springman, S. Hydro-Mechanical Analysis of a Surficial Landslide Triggered by Artificial Rainfall: The Ruedlingen Field Experiment. *Géotechnique* **2021**, *71*, 96–109. [CrossRef]
10. Bishop, A.W.; Blight, G.E. Some Aspects of Effective Stress in Saturated and Partly Saturated Soils. *Geotechnique* **1963**, *13*, 177–197. [CrossRef]
11. Fredlund, D.G.; Morgenstern, N.R. Stress State Variables for Unsaturated Soils. *J. Geotech. Geoenviron. Eng.* **1977**, *103*, 447–466.
12. Gens, A. Soil–Environment Interactions in Geotechnical Engineering. *Géotechnique* **2010**, *60*, 3–74. [CrossRef]
13. Alonso, E.E.; Pereira, J.M.; Vaunat, J.; Olivella, S. A Microstructurally Based Effective Stress for Unsaturated Soils. *Geotechnique* **2010**, *60*, 913–925. [CrossRef]
14. Tarantino, A. A Possible Critical State Framework for Unsaturated Compacted Soils. *Geotechnique* **2007**, *57*, 385–389. [CrossRef]
15. Fredlund, D.G.; Xing, A.; Fredlund, M.D.; Barbour, S.L. The Relationship of the Unsaturated Soil Shear Strength to the Soil-Water Characteristic Curve. *Can. Geotech. J.* **1996**, *33*, 440–448. [CrossRef]
16. Vanapalli, S.K.; Fredlund, D.G.; Pufahl, D.E.; Clifton, A.W. Model for the Prediction of Shear Strength with Respect to Soil Suction. *Can. Geotech. J.* **1996**, *33*, 379–392. [CrossRef]
17. Al-Sharrad, M.A.; Gallipoli, D.; Wheeler, S.J. Experimental Investigation of Evolving Anisotropy in Unsaturated Soils. *Geotechnique* **2017**, *67*, 1033–1049. [CrossRef]
18. Askarinejad, A. Failure Mechanisms in Unsaturated Silty Sand Slopes Triggered by Rainfall. 2015. Available online: https://books.google.com.hk/books?hl=en&lr=&id=dlsBCwAAQBAJ&oi=fnd&pg=PP1&dq=Failure+Mechanisms+in+Unsaturated+Silty+Sand+Slopes+Triggered+by+Rainfall&ots=Lfw48yZeTp&sig=g36q8g2JoyVjJ_IA8Z0w4NLDq-c&redir_esc=y#v=onepage&q=Failure%20Mechanisms%20in%20Unsaturated%20Silty%20Sand%20Slopes%20Triggered%20by%20Rainfall&f=false (accessed on 27 January 2021).
19. Casini, F.; Jommi, C.; Springman, S. A Laboratory Investigation on an Undisturbed Silty Sand from a Slope Prone to Landsliding. *Granul. Matter* **2010**, *12*, 303–316. [CrossRef]
20. Brönnimann, C.; Tacher, L.; Askarinejad, A.; Kienzler, P.; Springman, S.M. Pore Water Pressure Modelling in a Rainfall Triggered Shallow Landslide. In Proceedings of the 7th Swiss Geoscience Meeting, Neuchâtel, Switzerland, 20–21 November 2009; pp. 280–281.
21. Springman, S.; Askarinejad, A.; Casini, F.; Friedel, S. Lesson Learnt from Field Tests in Some Potentially Unstable Slopes in Switzerland. *Acta Geotech. Slov.* **2012**, *1*, 5–29.
22. Askarinejad, A.; Casini, F.; Kienzler, P.; Teysseire, P.; Springman, S.M. Mountain Risks: Two Case Histories of Landslides Induced by Artificial Rainfall on Steep Slopes. In *Mountain Risks: Bringing Science to Society, Proceedings of the "Mountain Risks" International Conference, Firenze, Italy, 24–26 November 2010*; CERG Editions; 2010; Available online: https://art.torvergata.it/handle/2108/93850#.YBE-mU_itPY (accessed on 27 January 2021).
23. van Genuchten, M.T. A Closed-Form Equation for Predicting the Hydraulic Conductivity of Unsaturated Soils. *Soil Sci. Soc. Am. J.* **1980**, *44*, 892–898. [CrossRef]

24. Casini, F. Deformation Induced by Wetting: A Simple Model. *Can. Geotech. J.* **2012**, *49*, 954–960. [CrossRef]
25. Fraccica, A.; Romero, E.; Fourcaud, T.; Sondon, M.; Gandarillas, L. Tensile Strength of a Vegetated and Partially Saturated Soil. *E3S Web Conf.* **2020**, *195*. [CrossRef]

Article

Experimental and Numerical Investigations of a River Embankment Model under Transient Seepage Conditions

Roberta Ventini [1,*], **Elena Dodaro** [2], **Carmine Gerardo Gragnano** [2], **Daniela Giretti** [3] **and Marianna Pirone** [1]

[1]   Department of Civil, Building and Environmental Engineering, University of Napoli Federico II, 80125 Napoli, Italy; marianna.pirone@unina.it

[2]   Department of Civil, Chemical, Environmental and Materials Engineering, University of Bologna, 40136 Bologna, Italy; elena.dodaro2@unibo.it (E.D.); carmine.gragnano2@unibo.it (C.G.G.)

[3]   ISMGEO srl, 24068 Seriate, Italy; giretti@ismgeo.it

*   Correspondence: roberta.ventini@unina.it; Tel.: +39-081-768-3617

Abstract: The evaluation of riverbank stability often represents an underrated problem in engineering practice, but is also a topical geotechnical research issue. In fact, it is certainly true that soil water content and pore water pressure distributions in the riverbank materials vary with time, due to the changeable effects of hydrometric and climatic boundary conditions, strongly influencing the bank stability conditions. Nonetheless, the assessment of hydraulic and mechanical behavior of embankments are currently performed under the simplified hypothesis of steady-state seepage, generally neglecting the unsaturated soil related issues. In this paper, a comprehensive procedure for properly defining the key aspects of the problem is presented and, in particular, the soil characterization in partially saturated conditions of a suitably compacted mixture of sand and finer material, typical of flood embankments of the main river Po tributaries (Italy), is reported. The laboratory results have then been considered for modelling the embankment performance under transient seepage and following a set of possible hydrometric peaks. The outcome of the present contribution may provide meaningful geotechnical insights, for practitioners and researchers, in the flood risk assessment of river embankments.

Keywords: riverbank; unsaturated soils; water retention curve; unsaturated permeability curve; transient seepage; slope stability

Citation: Ventini, R.; Dodaro, E.; Gragnano, C.G.; Giretti, D.; Pirone, M. Experimental and Numerical Investigations of a River Embankment Model under Transient Seepage Conditions. *Geosciences* **2021**, *11*, 192. https://doi.org/10.3390/geosciences11050192

Academic Editors: Jesus Martinez-Frias, Roberto Vassallo, Luca Comegna and Roberto Valentino

Received: 31 March 2021
Accepted: 25 April 2021
Published: 29 April 2021

Publisher's Note: MDPI stays neutral with regard to jurisdictional claims in published maps and institutional affiliations.

1. Introduction

Riverbanks are passive defense works, consisting of earthen structures built along the edges of a stream or river channel to prevent flooding of the adjacent land. In recent years, the risk assessment of river embankments has received great attention as consequence of the considerable socio-economic damages caused by flooding. Since 2000, floods in Europe have caused at least 700 deaths, the evacuation of about half a million people, and at least 25 billion euro in insured economic losses [1]. The economic impact of climate-related events varies considerably across countries. In absolute terms, the highest economic losses in the period 1980–2019 were registered in Germany, followed by Italy, then France [2].

Nevertheless, investments in earth retaining structures are relatively limited compared to other hydraulic engineering works, such as water regulation and canalization. In many countries, including Italy, minor embankments are not even subjected to any formal legislation or published technical guidelines and this has a severe impact on the performance of levees and safety of territories [3].

The design of riverbanks and their individual components depends on many factors, including functionality under different hydraulic loadings, feasibility, but also environmental constraint and availability of the materials. Numerous cross-sectional variations exist, each with the primary objective of reducing flood risk within the surrounding area. The simplest type of section is the homogeneous earthfill, composed of granular and cohesive soil, obtained from site excavation or borrow sources, though riverbank may also present an impermeable core (zoned embankment) or be composite with superstructures, waterside or landside berm and internal structures [4].

Riverbank failure, defined as the inability to prevent the inundation of the area surrounding a watercourse, may occur due to various hydraulic and structural causes, such as overtopping, seepage flow through foundation layer, internal and external erosion, slopes instability, scour and liquefaction [5]. Some possible failure mechanisms are reported in Figure 1. One of the most critical is represented by the overall macro-instability of the slopes triggered by the unfavorable seepage of water through the riverbanks, due to rivers persistently running at unusually high levels. An increasing number of studies have recognized changes in pore water pressures within a riverbank as a fundamental factor in determining conditions of instability and in the triggering of bank failures [6–9]. Moreover, burrowing animals are acknowledged by agencies responsible for earthen dams and riverbanks to have a significant role in substantial and costly damages [10].

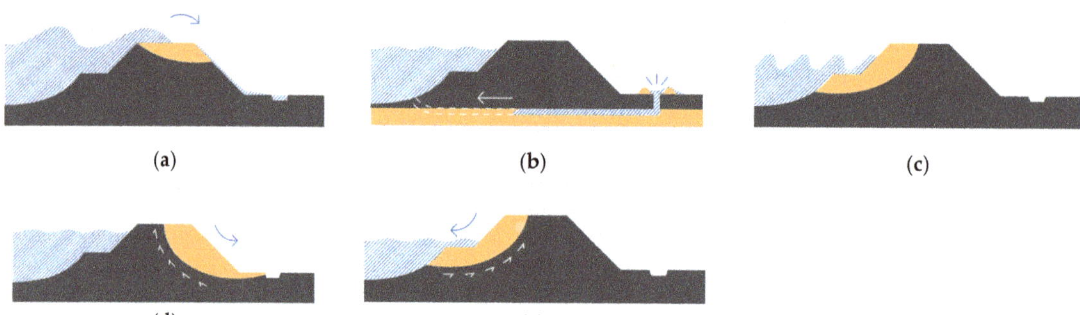

Figure 1. Overview of main riverbank failure mechanisms in earthen structures, modified from Kok et al. [11]: (**a**) Overtopping; (**b**) Uplift and piping; (**c**) Erosion of waterside slope; (**d**) Macro-instability in landside slope; (**e**) Macro-instability in waterside slope.

The development of high pore water pressures, due to a persistent groundwater flow, actually leads to a decrease in the soil shear resistance within the riverbank that may result in catastrophic slope failure on the landside [12,13]. This type of collapse strongly depends on the retention characteristics of the filling material such as suction variations with time, also due to seasonal variations of the meteorological conditions. In fact, the recurring soil water content variations in the embankment induced by hydrometric fluctuations and rainfalls occurrences may dramatically reduce the suction contribution to shear strength and eventually modify the microstructure of the soil, even creating a cracking state in the riverbank that could form preferential flow channels with general increase of the soil permeability and decrease of the shear strength [14]. As an example, the study of the breach that occurred along the river Secchia, Northern Italy, on 19 January 2014, provides clear evidence of a landside slope failure, caused by a series of simultaneous river stage raising combined to rainfall on the embankment surface and the wildlife activity along fluvial systems [15]. For the failure of the landslide, it is well known that toe drainage by gravelly materials is able to effectively lower the phreatic surface and to prevent water from seeping out of the 'landslide slope'.

The failure of the riverside is relatively different and could be triggered by external erosion of the toe and foundation of the embankment, primarily due to the action of waves, currents or turbulence; filling material is so removed from the waterside surface, leading to loss of stability and consequently to deep or shallow slope failure. A recurrent phenomenon causing riverside slope failure is also the rapid drawdown, typically occurring when the hydrometric level undergoes a rapid decrease, after a high stage, as the bank material is still near a saturated condition and the confining pressure of the river decreases to zero [8,16,17]. A frequently adopted solution for averting the aforesaid failure generally consists in protecting the riverside slope of the embankment using armourstone, gabions and concrete slabs. In general, different types of interventions, including the realization of components, such as impervious layers, cut-offs barriers, drains, and reinforcement layers within the embankment sections, are indeed considered by management and streambank control agencies to improve flood protection [4].

The identification of the main factors adversely affecting riverbank stability and triggering failure conditions is still a fundamental step for the optimal design and verification of these linear infrastructures, together with the individuation of those critical sectors, where structural improvements are mostly needed. The hydraulic and mechanical behavior of a riverbank during high water events can be generally investigated by a suitable combination of transient seepage and stability analyses, taking into account appropriate unsaturated soil geotechnical characteristics. In this framework, the assessment of hydromechanical parameters of the soil constituting the body of the embankment and foundation in partial and total saturated conditions is necessary in order to carry out proper predictive analyses of the collapse mechanism.

This paper describes a simple procedure for experimentally estimating optimal soil properties for riverbanks and their implementation in numerical analyses. The study case and materials are selected as representative for the riverbank systems of the alpine and Apennines tributaries of river Po (generally 5–10 m high above the ground level), which recently experienced various sudden overall collapses (Figure 2). Results hereafter discussed, besides constitutes a preliminary benchmark for the design of physical models that, in further studies, will be tested with the final aim to develop diagnostic and operational tools for the assessment of river embankments resistance to flood waves and support local governments to draw their resilience action plans.

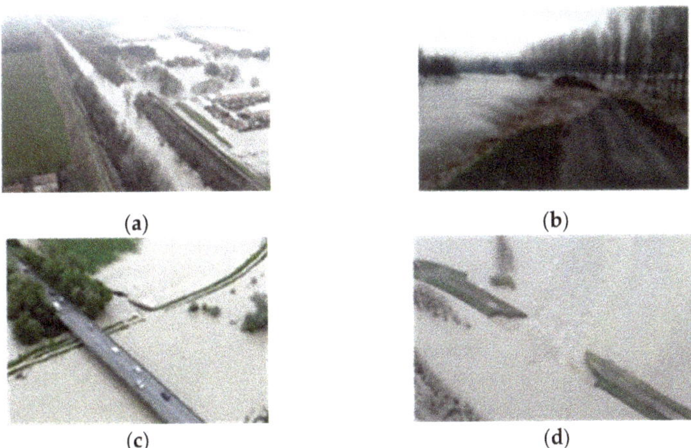

(a)   (b)
(c)   (d)

Figure 2. Recent riverbank breaches occurred in: (**a**) San Matteo (MO), Secchia river, 2014; (**b**) Lentigione (RE), Enza river, 2017; (**c**) Villafranca (FC), Montone river, 2019; (**d**) Nonantola (MO), Panaro river, 2020.

2. Materials and Methods

2.1. Case Study

The geometry considered in present study has been defined as suitable for replicating a typical failure for the embankments of tributaries of the Po river and concurrently reproducible in scaled physical models. The case study is thus referred to a simplified geometry (sketched in Figure 3) where the riverside and landside are equally sloped (45°);

the height of the embankment, which has a simple trapezoidal shape, is 7.5 m above the surrounding level and the crown is 3.0 m wide.

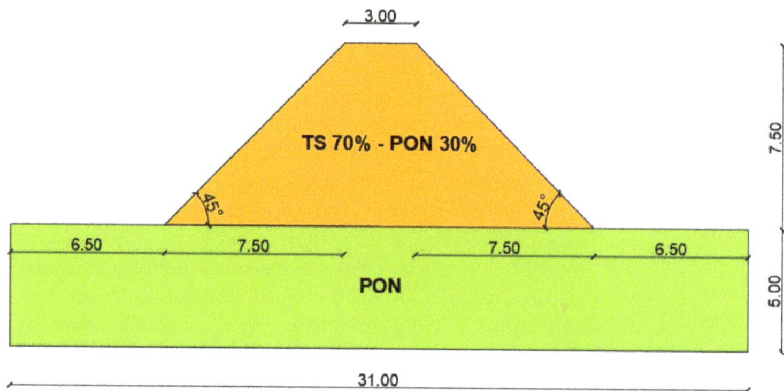

Figure 3. Geometry of the section of the riverbank analyzed.

The embankments of tributaries of river Po are prevalently constituted by a heterogeneous mixture of sands and silt, sometimes clayey soils in the above ground part, while the subsoil frequently consists of clayey and silty deposits of floodplain environment. Similarly, the soil here selected for the earthfill is constituted by a compacted mixture of 70% Ticino sand (TS) and 30% Pontida clay (PON), while a foundation characterized by a homogeneous consolidated layer of Pontida clay is taken as reference.

The whole model is supposed to be contained in a rigid box and this leads to peculiar boundary conditions for the foundation layer, which will be detailed in Section 4.1.1.

Particular attention will be devoted to riverside slope instability due to rapid drawdown after high water stages and to the landside instability, due to a consistent saturation of the embankment body.

2.2. Soil Physical Properties

Ticino sand (TS) is a well-known Italian sand which has been subjected to intensive conventional and advanced laboratory tests [18–22]. It is a uniform coarse to medium sand made of angular to sub rounded particles and composed of 30% quartz, 65% feldspar and 5% mica [18–23]. Pontida Clay (PON) is a low plasticity kaolinitic clayey silt obtained from a quarry of fine material located in Pontida (Italy), deposited in a post-glacial lake environment. X-ray diffraction tests indicate the following mineralogical composition: 25–30% illite, 20–25% kaolinite, 20–25% quartz, 12% calcite and dolomite, less than 10% feldspar, and less than 5% chlorite.

The grain size distributions for each single component (TS and PON) and the considered mixture (TS70%-PON30%) are given in Figure 4. Following a preliminary stage of laboratory investigations, soil index and physical main properties have been determined and are listed in Table 1.

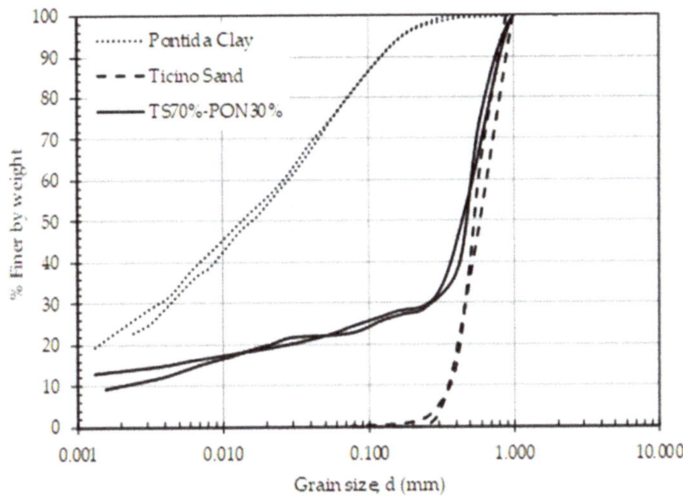

Figure 4. Grain size distributions of the tested materials.

Table 1. Average on index properties of the tested materials (γ_{min} maximum value of soil unit weight, γ_{max} minimum value of soil unit weight, e_{min} minimum value of void ratio, e_{max} maximum value of void ratio, G_s specific soil density, D_{50} mean particle size, U_c uniformity coefficient, LL limit liquid, PL plastic limit, PI plasticity index).

Soil	γ_{min} kN/m^3	γ_{max} kN/m^3	e_{min} -	e_{max} -	G_s -	D_{50} mm	U_c -	LL %	PL %	PI %
TS	13.64	16.67	0.574	0.923	2.671 [1]	0.574 [1]	1.83 [1]	-	-	-
PON	-	-	-	-	2.744 [2]	0.015 [1]	-	23.61 [1]	13.13 [1]	10.48 [1]
TS70%-PON30%	13.48	21.30	0.236	0.953	2.684 [2]	0.458 [1]	246.06	17.66	10.23	7.42

[1] Average of two measurements. [2] Average of three measurements.

2.3. Specimen Preparation and Initial Condition

Soils commonly used as embankment fill are compacted to a dense state to obtain satisfactory engineering properties such as shear strength and/or permeability. The degree of compaction and the molding water content needed to achieve the required compaction degree can be determined through standard laboratory tests. Above all, the optimum water content and maximum dry unit weight values of a soil can be determined under standard Proctor energy (SPE) and modified Proctor energy (MPE), which represent a quite reproducible compaction procedure and typical for embankments and retaining structures. The SPE allows to obtain a degree of compaction substantially lower than the modified maximum dry unit weight. Since more energy is applied for compaction using MPE, the soil particles are more closely packed and the optimum moisture content is lower. Higher energy results in greater shear strength, lower air voids and decreased permeability. Here, the modified Proctor compaction test has been performed on TS70%-PON30% following ASTM standard [24] using an automatic compactor. The specimen properties so obtained are reported in Table 2.

Table 2. Index properties and initial suction of the TS70%-PON30% specimens compacted at the optimum Modified Proctor (γ_d dry unit weight, w natural water content, n soil porosity, e void ratio, S_r degree of saturation).

Specimen	γ_d kN/m^3	w %	n -	e -	S_r %	Suction kPa
MP1	20.89	8.20	0.207	0.261	84.47	-
MP2	21.24	7.39	0.193	0.240	82.72	-
MP3	20.79	7.80	0.211	0.267	78.53	4.5
MP4	21.10	7.38	0.199	0.248	79.87	8
MP5	20.82	7.30	0.209	0.265	74.00	6
Average	20.97	7.61	0.204	0.256	79.92	6.17
Standard Deviation	0.17	0.34	0.01	0.01	3.62	1.43

Compacted soils used in earth structures are often in unsaturated conditions. Since the Modified Proctor compaction test reproduces the staged construction of earth embankments that is expected to be in unsaturated conditions, thus, immediately after this test the matric suction was measured in order to check the initial conditions (Table 2).

After instrument saturation, the porous stone of a small-tip tensiometer (Figure 5a) was coated with a slurry of Pontida clay, to provide adhesion between the ceramic cup and the soil in which it was inserted. Then, with the help of a screw of the same diameter of the porous cup, a hole was created into the soil contained into the Proctor mold. The bottom of the hole laid at a distance of 5 cm from the surface of the material, then, when the porous cup was finally inserted into the soil, the superficial and disturbed layer was removed and the measurement returned a suction value that can be assumed representative of the entire compacted material contained into the mold. When the hole was excavated and the porous cup inserted, the sample and the small-tip tensiometer were hermetically covered, to avoid any input of air that would compromise the accuracy of the measurements. Considering the grain size distributions involved and the type of the instrument used, the equilibrium was reached in few hours within 100 and 200 min. The tests have been carried out with two small-tip tensiometers to verify the reliability of the measurements. At the equilibrium, the two instruments gave two overlapping trends of the suction values for all the specimens, as shown in Figure 5b. The measured suction values are in general low, ranging between 4.5 kPa and 8 kPa, mainly due to high percentage of TS in the mixture.

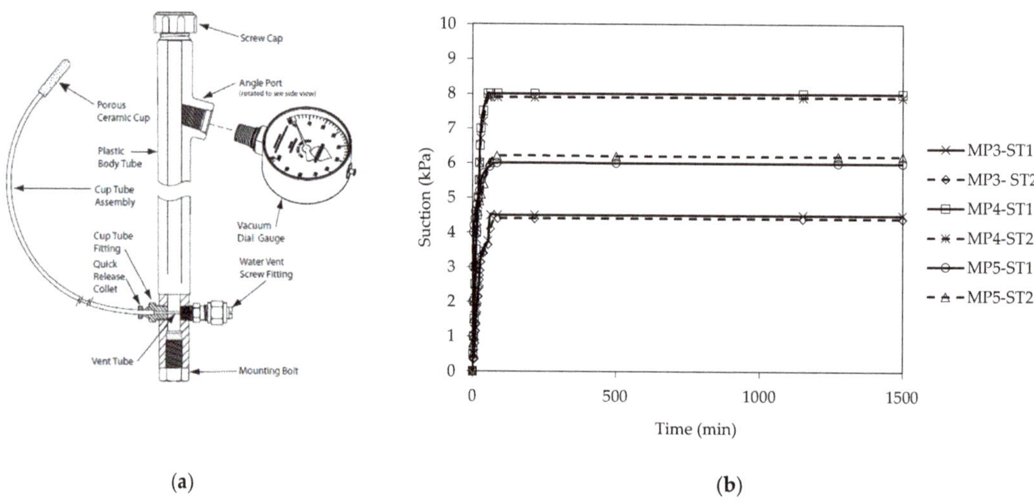

(a)   (b)

Figure 5. Suction measurements in samples after compaction: (**a**) Soil moisture Probe Parts; (**b**) Matric suction vs. Time.

3. Soil Hydro-Mechanical Characterization

3.1. TS70%-PON30%

3.1.1. Saturated Permeability from Permeameter Test

The saturated permeability was determined through a constant head test in a permeameter by measuring the volumes of water exchanged by the saturated specimen. This test relies on the Darcy's law [25], reported in Equation (1):

$$\Delta V / (i \cdot A) = k_{sat} \cdot \Delta t \tag{1}$$

where k_{sat} is the saturated permeability, A is the cross-section area of the specimen, ΔV is the volume of water that flowed through the specimen in the time interval Δt, and i is the hydraulic gradient. The volume ΔV is equal to the average of the volume of water flowing inside and outside of the specimen for each time step.

The test was performed on saturated soil specimens using the setup presented in [26]. The equipment is composed of a permeameter in which the sample is placed, two graduated burettes, two water reservoirs, two air pressure regulators and a pressure transducer. Two porous plates (previously saturated) are placed on the top and bottom of the specimen, connecting the soil water to the drainage grooves machined in the sealing acrylic caps. Two O-rings are used to seal the caps to prevent any leakage of water. For the same reason, the acrylic caps are tied together with a well tightened screw system. Finally, filter papers for sandy soils are placed on top and bottom of the specimen to avoid the blockage of the porous plates.

The soil specimen was taken from the Proctor mold immediately after suction measurements were carried out. The specimen is contained in a cylindrical steel ring 7.21 cm in diameter and 6.08 cm in height (total volume of about 248.7 cm^3). The saturation is obtained by flushing distilled and deaerated water into the specimen. Then, a small and constant pressure head gradient (5 kPa between bottom and top) is applied to the specimens to induce a water flow in the upward direction. The saturation process is controlled by continuously monitoring the volumes of water coming in and out of the specimen; it is assumed that the saturation process is complete when a stationary condition is reached, and the incoming and outgoing flow rates become equal [26]. In order to quantify k_{sat}, the variable E' has been plotted against the time (Figure 6), with E' defined as the volume of water discharge per unit bulk cross sectional area and hydraulic gradient (Equation (2) [25]):

$$E' = \Delta V / (i \cdot A) = k_{sat} \cdot t \tag{2}$$

where k_{sat} is equal to the slope of the linear regression in the t–E' space.

Since this type of test could be affected by random errors, due to the operator's readiness during the readings from two burettes at a time, four measurements have been made for each of the four compacted specimens, as reported in Figure 6.

The values determined in permeameter evidence a strong variability in results, despite of similar porosities. Some dispersion can be surely related to heterogeneity that can occur during the compaction among all the sampled specimens. The saturated permeability values determined for each specimen of TS70%-PON30% mixture are reported in Table 3, with the mean logarithmic (of base 10) value equal to 1.29×10^{-7} m/s.

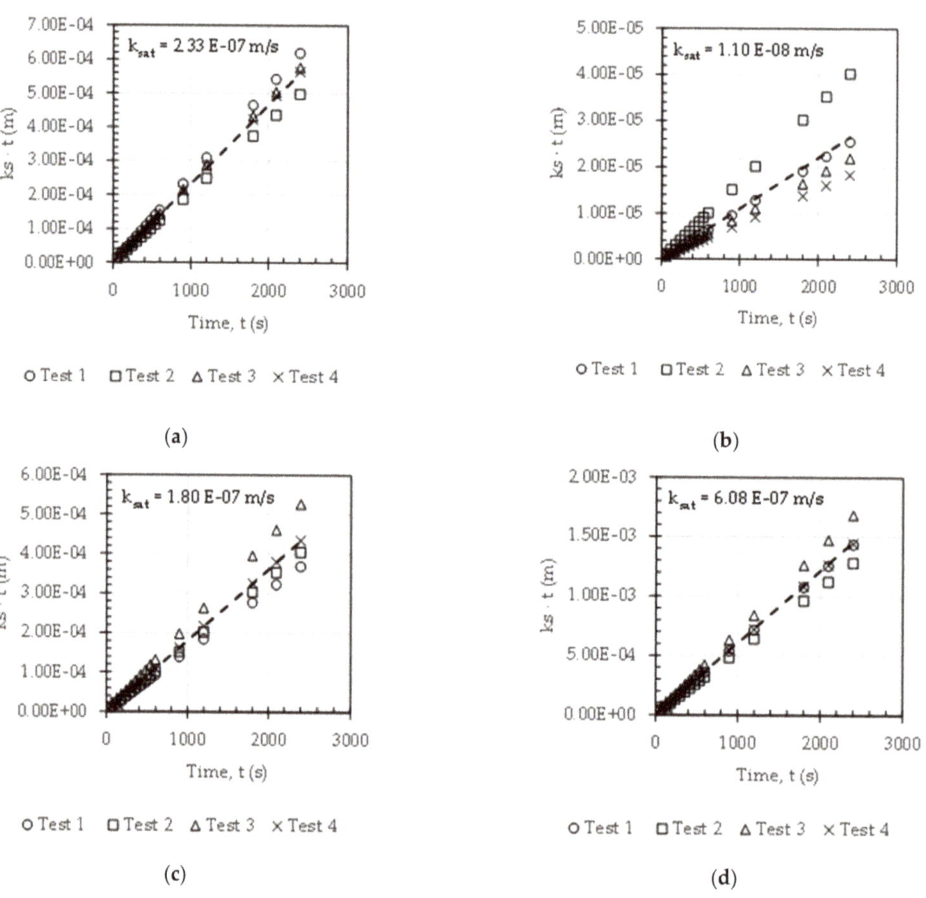

Figure 6. Saturated permeability measurements: (**a**) MP2; (**b**) MP3; (**c**) MP4; (**d**) MP5 specimens.

Table 3. Saturated permeability (k_{sat}) of the TS70%-PON30% specimens and hydraulic parameters of the Mualem–van Genuchten (MVG) retention model.

Specimen	k_{sat} m/s	θ_r -	θ_{sat} -	α 1/kPa	n_V -	l -	R^2 -
MP2	2.33×10^{-7}	-	-	-	-	-	-
MP3	1.10×10^{-8}	0.005	0.190	0.013	1.650	−0.030	0.920
MP4	1.80×10^{-7}	-	-	-	-	-	-
MP5	6.08×10^{-7}	0.017	0.200	0.011	1.400	−0.330	0.930
Logarithmic Mean	1.29×10^{-7}	0.011	0.195	0.012	1.525	−0.180	0.925
Standard Deviation	-	0.006	0.005	0.001	0.125	0.150	0.005

3.1.2. Soil Water Retention Curve (SWRC) and Hydraulic Conductivity Function (HCF)

In this study, the device described by [26] was used to identify the SWRC of the mixture along the main drying branch (Figure 7).

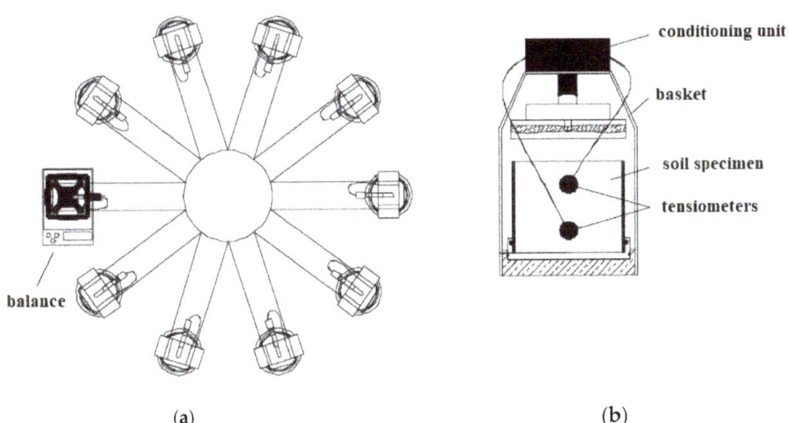

Figure 7. Apparatus for the evaporation tests, modified from Nicotera et al. [26]: (**a**) sampler changer; (**b**) specimen ring and sampler holder.

A commercial apparatus (ku-pF by Umwelt-Geräte-Technik GmbH) developed for performing evaporation tests according to the simplified procedure of [27] was adopted. After completely saturating the specimen in the permeameter, it was mounted in the ku-pF apparatus (Figure 7a) for carrying out the evaporation test. The specimen has been enclosed in a metallic ring which includes two holes to house the tensiometers installed 3 cm apart (respectively 1.5 and 4.5 cm from the top). Each pair of pressure sensors is connected to a conditioning unit arranged upon each sample holder (Figure 7b) that is fitted with an electronic assembly that digitalizes the tensiometer readings at selected intervals and sends them to the data logger via a two-wire line. The connection to the data logger takes place through contacts in the support rods on the rotating carrier.

The tensiometers were saturated before each test and the calibration checked. Once calibrated, the tensiometers were installed in the holes of the saturated specimen, previously made with a special guide. The material resulting from the drilling was kept aside, weighted, and then put into the stove for the evaluation of water content. Then, the specimen enclosed in the ring was placed on a perforated base covered with filter paper while the top was covered with clingfilm and a metal cup. At this point, the specimen was placed on the closed base of the basket of the ku-pF apparatus. The basket was installed in a rotating changer arm and was weighed every 10 min on a precision scale with a resolution of 0.01 g to register variations in water content.

At first, the test was conducted with top clingfilm and top cap still mounted on the specimen to reach the hydrostatic condition inside the specimen. The equilibrium was reached when the difference between the readings of top and bottom tensiometers was 0.3 kPa (the tensiometers are placed 3 cm apart). Then, when the equilibrium is reached, top clingfilm and top cap are removed and weighted. The actual evaporation process began and lasted until the tensiometers reached the cavitation value (about 80 kPa). During the evaporation test, the water content decrease of the sample was measured at set intervals by a digital balance logged into an acquisition system. After the evaporation test, the soil specimen was moved in the Pressure Plate apparatus [28,29] to determine additional experimental points for suction values between 90 kPa and 1 MPa. The results of tests on MP3 and MP5 specimens in the ku-pF apparatus are presented in Figure 8a on the soil water retention plane. In particular, the mean measured matric suction has been coupled to the average water content estimated by the measured variation in the soil weight during the evaporation test. On the same figure the points determined in the Pressure Plate at higher suction are overlapped.

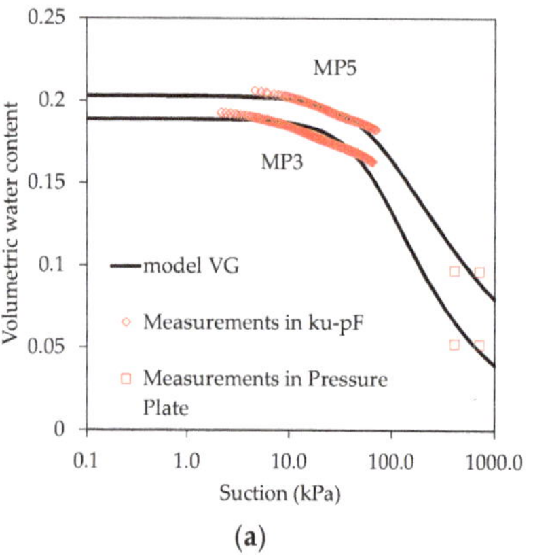
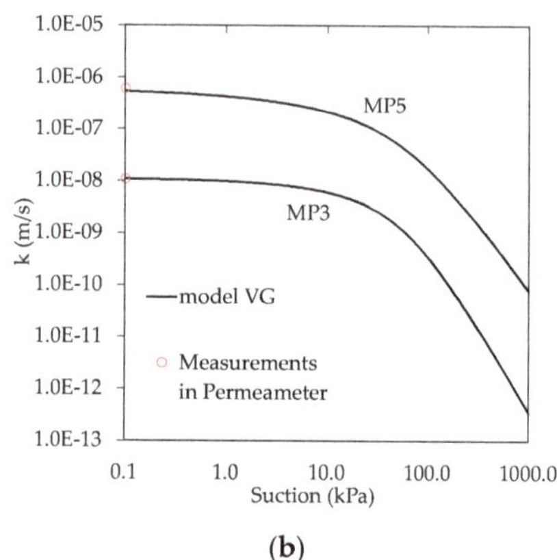

Figure 8. (a) **Soil Water Retention Curves** (SWRCs) and (b) **Hydraulic Conductivity Functions** (HCFs) determined by inverse analyses in Hydrus 1D (black continuous line) overlapped on measurements obtained in the ku-pF apparatus, in the Pressure Plate and in the permeameter (red symbols).

The Mualem–van Genuchten (MVG) model [30,31] is then considered in the interpretation of laboratory data through the software HYDRUS-1D [32], which numerically solves the Richards' equation using standard Galerkin-type linear finite element schemes. HYDRUS-1D is able to simulate the water movement in one-dimensional variably saturated media.

The relationship between the volumetric water content (θ) and the matric suction (s) in the MVG model is here described by Equation (3) [31], the relationship between the effective degree of saturation (S_e) (Equation (4) [31]), and the hydraulic conductivity (k) is described by Equation (5) [30]:

$$\theta = \theta_r + (\theta_{sat} - \theta_r) \cdot \left\{ \frac{1}{\left[1 + (\alpha \cdot s)^{n_v} \right]} \right\}^m \tag{3}$$

$$S_e = \frac{\theta - \theta_r}{\theta_{sat} - \theta_r} \tag{4}$$

$$k = k_{sat} \cdot S_e^{\,l} \cdot \left[1 - \left(1 - S_s^{\,1/m} \right)^m \right]^2 \tag{5}$$

where θ_r is the residual volumetric water content and θ_{sat} represents the volumetric water content at saturation; α, n_v and l are fitting parameters of the models while $m = 1 - 1/n_v$; k_{sat} is the saturated permeability and S_e is the effective degree of saturation.

The model parameters were estimated via inverse analysis of the experimental measurements by means HYDRUS-1D [32]. According to Nicotera et al. [26], application of the inverse method consists of a numerical analysis simulating a lab test, and of determination of the values of the model parameters for which differences between observed and simulated flow variables are minimized; the estimated values of the parameters are those that optimize the model response. Therefore, the best-fitting parameters of the MVG model were obtained by means of inverse analysis of the evaporation tests carried out in ku-pF apparatus. The parameter vector associated with the main drying (θ_d, θ_{sat}, α, n, l) was determined using the software HYDRUS-1D [32], in which the objective function used to fit the data was minimized using the Levenberg–Marquardt nonlinear minimization method [33,34]. The initial condition was fixed as the hydrostatic pressure distribution estimated from initial suction measurements. The water flow occurring within the specimen was vertical, with a null flux at the bottom and an upward flux at the upper boundary equal to the weight change in the specimen in the given time range. Data sets and respective weights in the objective function were composed of (i) matric suction measurements at the tensiometers during the monitoring time, with a weight of 1; (ii) the pair (s, θ) obtained from the pressure plate, with a weight of 1; and (iii) the pair (s, θ) corresponding to the AEV, with a weight of 1. The AEV was roughly identified as the point of maximum curvature on the SWRC curve, obtained by coupling the mean measured matric suction to the average water content estimated by the measured variation in the soil weight during the evaporation test.

Suction and water content numerically determined are overlapped to the experimental results in Figure 8a for tests on MP3 and on MP5 specimens. In Figure 8b, the permeability curve is shown where the value of saturated permeability measured is also overlapped. The fitting parameters of the MVG model determined by inverse analyses for each sample are reported in Table 3.

Lastly, it is important to keep in mind that, in this study, the hydraulic hysteresis is neglected and the hydraulic soil behavior has been modelled using a single branch (drying) of the SWRC. According to the literature [34,35], the main wetting branch and the scanning paths are located below the main drying curve on the SWR plane, these are characterized by hydraulic conductivities generally lower than those detected along drying curves. Therefore, neglecting hysteresis could cause errors in the mass flux calculations and soil shear strength determination.

3.1.3. Mechanical Soil Properties at Full Saturation

Monotonic triaxial compression tests were performed to study the stress–strain-strength behavior of the TS70%-PON30% mixture. Six monotonic consolidated drained (CD) triaxial tests were carried out using different confining pressures (σ'_c) on TS70%-PON30% specimens with 38 mm diameter and 76 mm height (Table 4). Specimens were compacted at the optimum modified Proctor (MPE). The weighed materials were moistened uniformly and then two modified Proctor compaction tests were done to prepare two sets of three specimens from each Proctor mold.

Table 4. Characteristics of TS70%-PON30% specimens used in triaxial testing.

Sample	Specimen	γ_d (kN/m^3)	w (%)	n (-)	σ'_c (kPa)
	MP6_1	20.84	7.92	0.210	50
MP6	MP6_2	21.01	7.54	0.203	100
	MP6_3	20.89	7.57	0.208	200
	MP7_1	21.02	7.39	0.203	75
MP7	MP7_2	21.07	7.40	0.201	150
	MP7_3	20.79	7.21	0.212	300
Average		20.94	7.51	0.206	
Standard Deviation		0.10	0.22	0.004	

All monotonic tests were performed in drained conditions (CD) using a strain-controlled compression loading system. The prepared specimens have been saturated by adopting base-top drainage technique in the triaxial apparatus until Skempton's B-value exceeded 0.98. After achieving full saturation, the specimens were subjected to the predefined confining pressures and then the axial load was applied up to an axial strain of about 20%. The imposed axial displacement rate was 0.05 mm per minute. During the tests, axial stresses, axial strains and volumetric strains were recorded.

Figure 9a,c show the variations of deviatoric stress and volumetric strain versus axial strain for tested material. As expected, the peak deviator stress increases with increasing applied confining pressure and the higher the value of confining pressure, the greater the axial strain corresponding at peak deviator stress. The rapid decrease in deviatoric stress after peak indicates the brittle response of the dense mixture. All specimens showed a dilating trend in their volume change, as evident also in Figure 9d. Only four specimens reached critical state condition (Figure 9a–c). The functions of the critical state line (CSL) in the q-p′ and e-p′ plane are also reported in Figure 9b,d while the peak strength envelope in the q-p′ plane is shown in Figure 9b.

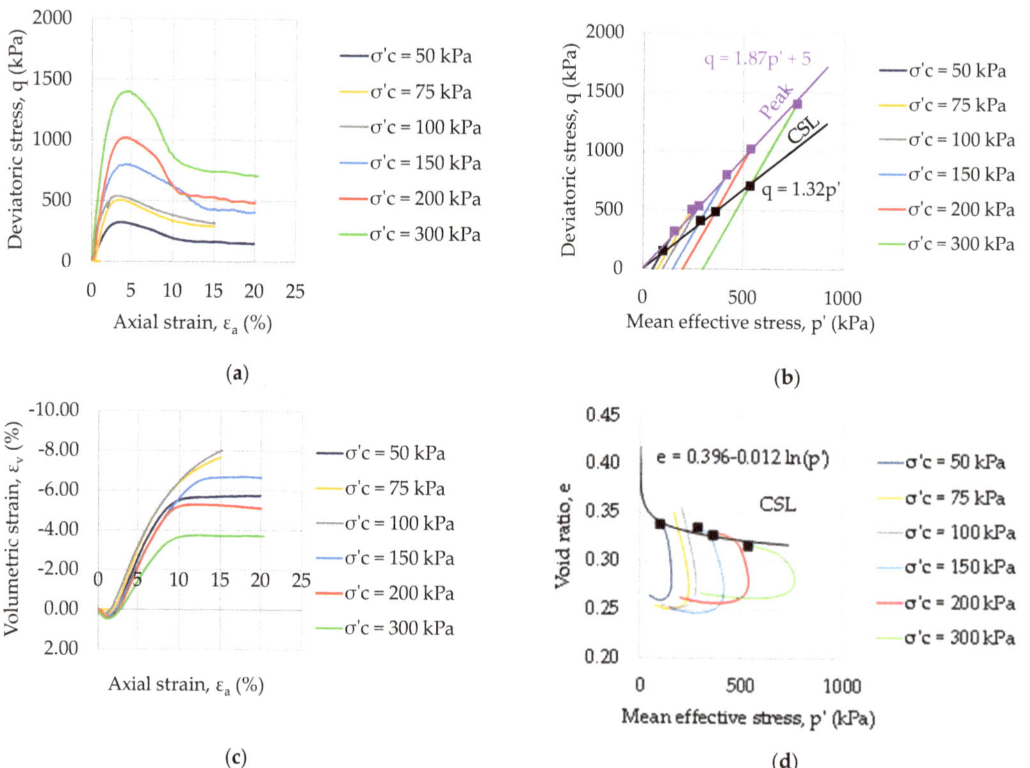

Figure 9. Stress-strain behavior of TS70%-PON30% mixture: (**a**) q-ε_a; (**b**) q-p′; (**c**) ε_v-ε_a; (**d**) e-p′ planes.

By processing the experimental data from CD triaxial tests, the angle of shearing resistance and cohesion values obtained are 45.5° and 5 kPa in peak condition and 34° at critical state condition, respectively.

3.2. Pontida Clay

Pontida Clay has been extensively used in the past to investigate the effect of temperature on the stress-strain behaviour of the clay skeleton and clay-water system [36,37].

To reproduce the state of a foundation layer deposited in a floodplain environment, a slurry of Pontida Clay was prepared at a water content w = 43%, equal to 1.8 time its limit liquid, (distilled water and dried Pontida clay were mixed under vacuum) and one-dimensionally compressed under a maximum vertical effective stress of 200 kPa.

Saturated permeability of Pontida Clay was derived from an oedometer test at different vertical effective stress as a function of the primary consolidation coefficient c_v and the compressibility coefficient m_v, according to Terzaghi's formulation of one-dimensional consolidation. Figure 10a,b show the computed permeability and the void ratio as a function of the applied vertical stress. Given the vertical overburden stress under the levee, ranging from about 100 to 150 kPa, a reference value $k_{sat} = 6.67 \times 10^{-10}$ m/s, relating to last loading test of the oedometer test, has been assumed for simulating the state of the levee foundation layer.

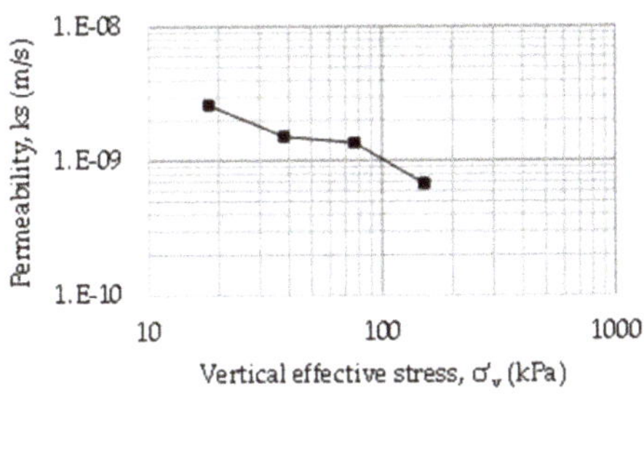
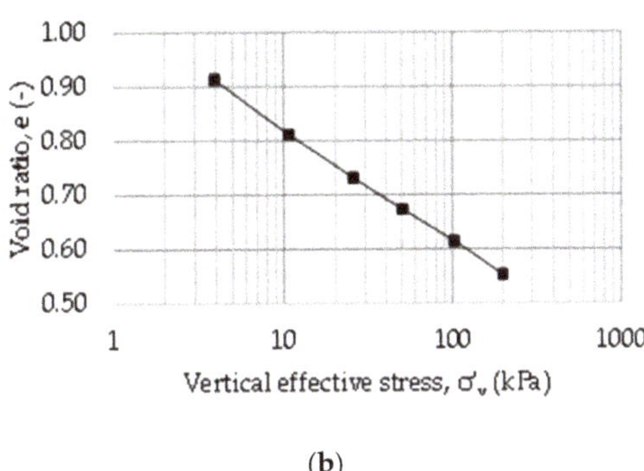

(a)   (b)

Figure 10. Pontida clay (PON): (**a**) saturated permeability and (**b**) oedometric curve, from oedometer test.

The mechanical properties of PON have been obtained through a series of drained and undrained triaxial tests carried out on both isotropically and anisotropically consolidated samples. The samples were obtained from large specimens, 1-D compressed from a slurry in a consolidometer. At the end of consolidation in the triaxial cell, some samples were normally consolidated, some others had an over consolidation ratio in the range 1.5–16. All the samples reached the critical state (or constant volume conditions). The state of the samples at critical state, evidenced with black squares in the q-p' (Figure 11a) and e-p' (Figure 11b) planes, have been interpreted to derive the critical state line (CSL), which has a slope M = 1.33 (i.e., a shearing resistance angle at critical state ϕ'_{cs} = 33°). The critical state conditions are also plotted in Figure 11b in the e-p' plane and have been interpreted via the logarithmic equation reported in the figure.

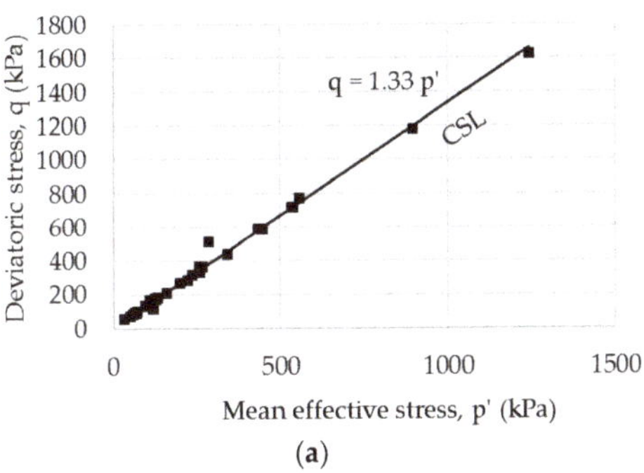
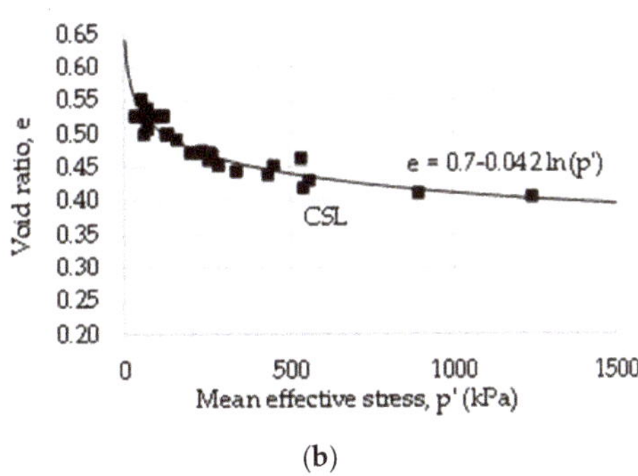

(a)   (b)

Figure 11. Pontida clay (PON) mechanical properties: critical state conditions in the (**a**) q-p' and (**b**) e-p' planes from triaxial tests.

4. Numerical Analysis on the Hydraulic Response and Stability Assessment

4.1. Seepage Analyses

The finite element software SEEP/W [38], capable of solving complex saturated and unsaturated groundwater flow problems, has been used to perform both steady-state and transient seepage analyses with reference to a bidimensional river embankment model, based on the simplified geometry sketched in Figure 3. The software calculation is formulated on the base of Darcy's law [25], originally derived for saturated soil. In the case of unsaturated porous media, the hydraulic conductivity is assumed to be a function of volumetric water content and matric suction [39].

The partial differential equation governing two-dimensional transient unsaturated water flow through riverbank is the Richard's mass conservation law, which states that the difference between the flux entering and leaving an elemental volume at any instant of time is equal to the change in storage of the system [39,40]. As implemented in the code, it is given by (Equation (6) [40]):

$$\frac{\partial}{\partial x}\left(k_x \frac{\partial h}{\partial x}\right) + \frac{\partial}{\partial y}\left(k_y \frac{\partial h}{\partial y}\right) + Q = \frac{\partial \theta}{\partial t} \tag{6}$$

where h is the total hydraulic head, k_x and k_y represent the hydraulic conductivities in the x and y direction respectively, Q is the sink term, θ is the volumetric water content and t is the time. For the present application, soil permeability has been assumed to be as isotropic. This model assumes that the soil porous medium is rigid under partially saturated conditions (no volume deformations, Equation (6)). However, SEEP/W is able to simulate bidimensional un-coupled consolidation in the fully saturated soil by specifying the coefficient of volume compressibility. When the hydraulic load is applied at soil surface, un-coupled approach could not result rigorous and a fully coupled stress/pore-water pressure analysis might be required. Especially in rapid drawdown phenomena, the un-coupled approach could overestimate pore water pressures respect to fully coupled analysis. This reflects in the prediction of lower safety factor, in favor of safety. Nevertheless, in this study, the coefficient of volume compressibility of mixture is set equal to zero; this hypothesis seems here reasonable due to the predominantly granular texture of the compacted mixture, which is characterized by significant stiffness. Therefore, the differences in terms of positive pore water pressures, and thus factor of safety, computed by the uncoupled approach, adopted here, and by the fully coupled one are expected to be small.

Furthermore, at the present stage of the analysis, soil-atmosphere interaction processes have been neglected in seepage modelling.

Under steady-state conditions, which may be reasonably referred to extended period of stationary river level, the water storage in the model is constant in time, so the right side of the differential Equation (6) reduces to zero.

An unstructured mesh, characterized by an irregular pattern, consisting of 5878 triangular and quadrangular elements, with an approximate element global size of 0.20 m, was adopted for the entire domain as shown in Figure 12. The mesh density was iteratively optimized by reducing the element size until no significant change in pore-water pressure was observed after refinement.

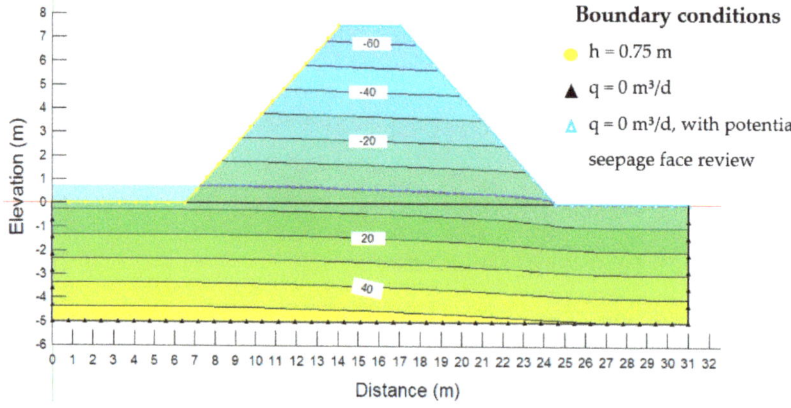

Figure 12. Initial pore water pressure distribution for transient seepage analysis (isoline increment = 10 kPa), boundary conditions and modelling mesh. Phreatic surface in blue bold line.

Soil permeability of the foundation layer was considered constant (only saturated conditions) and equal to its saturated value, while the soil forming the embankment was assumed as partially saturated considering the average parameters listed in Table 3 and Equations (3) and (5) as retention and hydraulic functions.

4.1.1. Boundary and Initial Conditions

To simulate a stationary flow through the embankment for a persistent retained high water level, a steady-state seepage analysis was performed assuming a maximum hydrometric height of 6.75 m (90% of the maximum embankment height) on the riverside, with the landside water table located at the ground level. As mentioned, since the model is supposed to be contained in a rigid box, the bottom horizontal, right, and left vertical outlines of the foundation layer were assumed to be impermeable. A no flow condition with a potential seepage face review has been assigned, instead, to the crest of the embankment and the landside slope; this implies that the water flow will remain zero until the pore water pressures computed by the code are not positive, otherwise the boundary condition will be reviewed to zero pressure head.

Concerning transient seepage analysis, the definition of initial condition in terms of suction and pore water pressure represents a crucial, preliminary, aspect. In the present study, a steady-state regime associated to a 0.75 m retained water level upon the riverside ground level and a hydrostatic suction pattern above the phreatic line, is assumed as starting condition. Suction distributions measured on site are generally different from the hydrostatic pattern, e.g., [9], especially during the wet period, nevertheless this hypothesis has been considered admissible for a preliminary model stability evaluation.

With regard to boundary conditions in transient seepage analysis, a variable hydrometric condition was imposed to model river level fluctuations on the surfaces potentially affected by water action. In order to simulate a realistic series of river flood events, data from the sudden collapse occurred on 19 January 2014 on river Secchia (Figure 2a) were taken as reference. In this particular case, the hydrograph recorded from 25 December 2013 to 19 January 2014 was characterized by a rapid series of three hydrometric peaks, which occurred in about one month [41]. Similarly, in the present analysis, a synthetic hydrograph is so characterized by a series of three hydrometric peaks. Starting from a river stage of 0.75 m in the first two days, each high water event reaches a maximum level of 6.75 m, raising in one day, then returning to the minimum water elevation in three days after a day of hydrometric peak persistence. The time between two following events is six days, while the final high-water level lasts for a longer period (two days). The water level was modelled as time-dependent boundary condition, expressed with linear variations of total hydraulic head (Figure 13) with a total elapsed time of 30 days. Moreover, rainfall infiltration fluxes have not been considered to specifically focus on the transient effects of hydrometric variations.

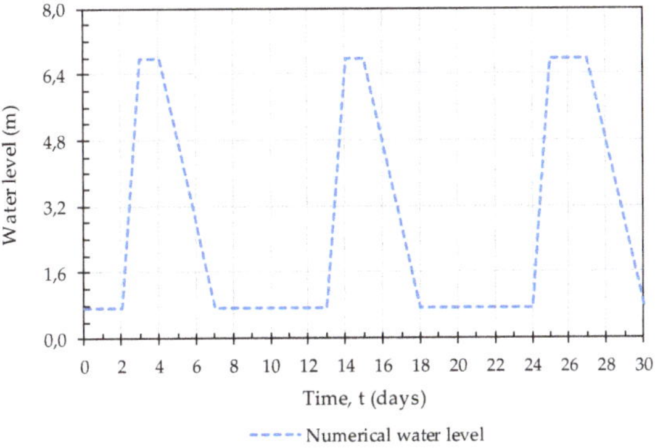

Figure 13. Numerical total hydraulic head boundary condition assumed in the transient seepage analysis.

4.2. Stability Analyses

The pore water pressure distributions obtained from seepage analysis, both in transient and steady state conditions, were used to evaluate the stability of the embankment on both riverside and landside slopes, expressed in terms of Factor of Safety. Stability analyses were performed with the software SLOPE/W [42] and adopting the Morgenstern and price limit equilibrium method [43].

Circular slip surfaces were generated by setting geometric constraints, consisting in the ranges for the entry and the exit zones of the sliding mechanisms, selected at the top and the toes of the embankment (Figure 14).

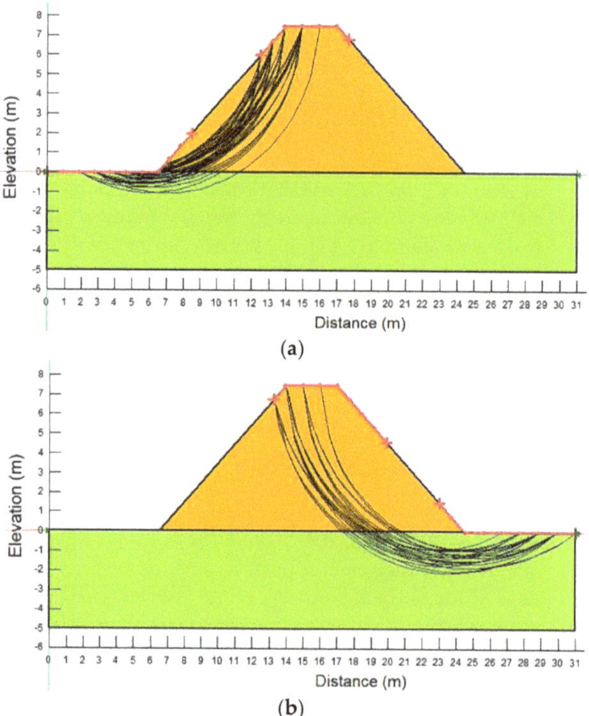

Figure 14. Slip surfaces investigated for the riverside (**a**) and the landside slopes (**b**). Entry and exit ranges of sliding mechanism are drawn in red.

The failure criterion of Vanapalli et al. [44], implemented in the software, was used for considering unsaturated shear strength contribution and it can be expressed in the form (Equation (7)):

$$\tau = c\prime + (\sigma_n - u_a)\tan \varphi\prime + (u_a - u_w)S_e\tan \varphi\prime \tag{7}$$

where τ is the shear strength, σ_n is the total normal stress, u_a is the pore air pressure, u_w is pore water pressure, c' is the effective cohesion, $\varphi\prime$ is the friction angle in terms of effective stress, and S_e is the effective degree of saturation. In saturated conditions, when the pore-air pressure, u_a, is equal to the pore water pressure, u_w, Equation (7) turns into the classical Mohr–Coulomb failure criterion and the shear strength is a function solely of the normal effective stress. The second part of the equation provides the shear strength contribution due to matric suction $(u_a - u_w)$, that is correlated to the soil water retention curve by the term S_e. The soil parameters adopted in the evaluation of the factors of safety, related to the sliding mechanisms hereafter investigated, are reported in Table 5.

Table 5. Soil parameters adopted in the river embankment stability analyses.

Material	γ_{sat} (kN/m^2)	φ' (°)	c' (kPa)
Embankment Body	22.96	45.50	5.00
Foundation Layer	21.14	33.00	2.00

4.3. Results and Discussion

The results of the steady-state analysis, performed taking the maximum hydrometric height, in terms of pore water pressure and total head distributions are showed in Figure 15. The phreatic surface reaches a considerable height on the landside slope (half of the overall embankment elevation) and the body of the embankment is almost fully saturated.

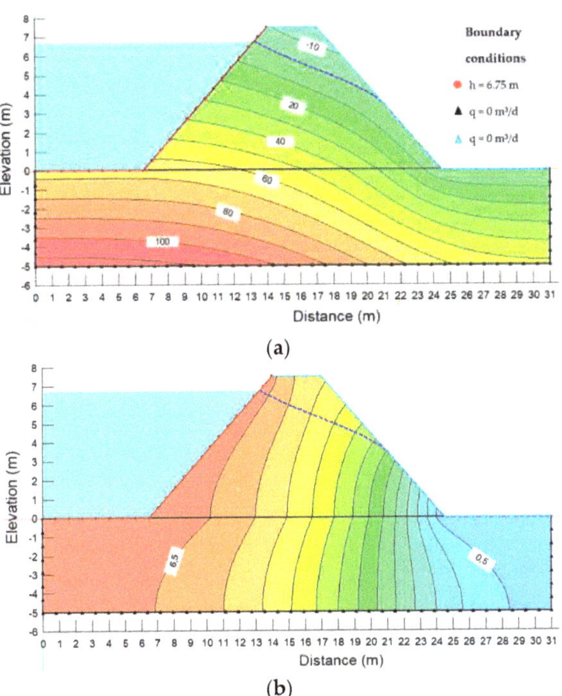

Figure 15. Results of the steady state seepage analysis: (**a**) pore water pressure distribution (isoline increment = 10 kPa) and (**b**) water total head distribution (isoline interval = 0.5 m). The blue bold line indicates the estimated location of the phreatic surface.

The factor of safety (FoS) associated to the steady-state regime investigated is just above one (1.05), which theoretically corresponds to an incipient condition of collapse. The fact that the FoS turned out to be slightly greater than one, despite the significant increase in pore water pressure and the substantial raise of the phreatic surface, is due to the high shear resistance of the TS70%-PON30% mixture, which constitutes the embankment body. It should be noted that the analyses in steady state seepage conditions, though not being representative for the specific case, may provide useful information with regard to the existing margin of safety of the river embankment, as the related FoS might be identified with the reference threshold for a limit condition. Nevertheless, only transient seepage analysis, taking into account the effective fluctuations of the river water level, can predict the actual safety conditions of an earthen retaining structure. Figure 16 summarizes the results of the transient seepage analysis in terms of pore water pressure contours, with reference to four significant time steps: (i) at the beginning of the first hydrometric peak (t = 3 days), (ii) after the first drawdown (t = 7 days), (iii) at the end of the second hydrometric peak (t = 15 days) and (iv) in correspondence of the end of the third peak (t = 27 days), while the hydrometric level is again at its maximum value.

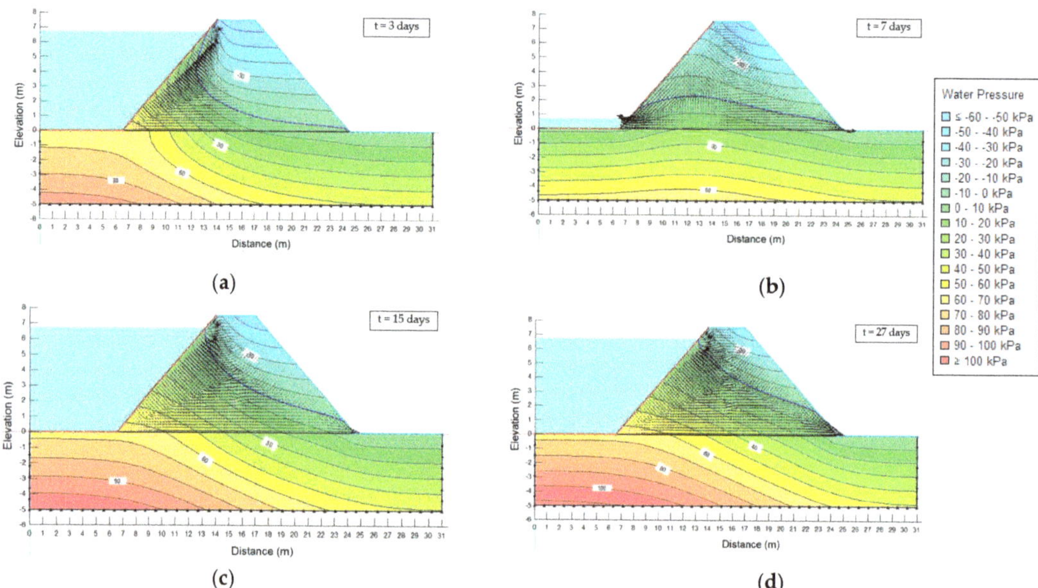

Figure 16. Pore water pressure distribution at different stages during the transient seepage analysis: (**a**) at the first peak of the hydrograph (t = 3 days), (**b**) after the first drawdown (t = 7 days), (**c**) during the second hydrometric peak (t = 15 days), (**d**) in correspondence of the end of the third peak (t = 27 days). The black arrows represent velocity vectors of the groundwater flow.

In the light of the transient seepage results plotted above, some meaningful observations can be made. The groundwater flow predominantly develops in the body embankment, due to the remarkable difference in hydraulic conductivities between this unit and the foundation layer (1.29×10^{-7} m/s and 6.67×10^{-10} m/s respectively). During the first high water event, about a third of the embankment turns out to be already in saturated conditions. This outcome can be fairly attributable to the retention properties, the significant hydraulic permeability which characterizes the TS70%-PON30% mixture, as well as to the magnitude of the flooding event on the embankment.

After the first drawdown, the phreatic surface within the riverbank does not come back to its initial configuration, on the contrary it remains at a considerable height in correspondence of the toe of the embankment on the riverside. It is worth noting that, while the drawdown occurs, the groundwater flow reverses its motion, heading from the bank towards the river. The progression of hydrometric peaks results in a gradual advancement of the phreatic surface towards the landside slope, which leads to a significant increasing of the pore water pressure in the entire riverbank section.

The correlation between river stages, the associated pore water pressures and their effects in terms of bank stability is evident observing the evolutions of the factor of safety over time, both for the riverside and landside slopes, presented in the following diagram together with the hydrograph considered (Figure 17).

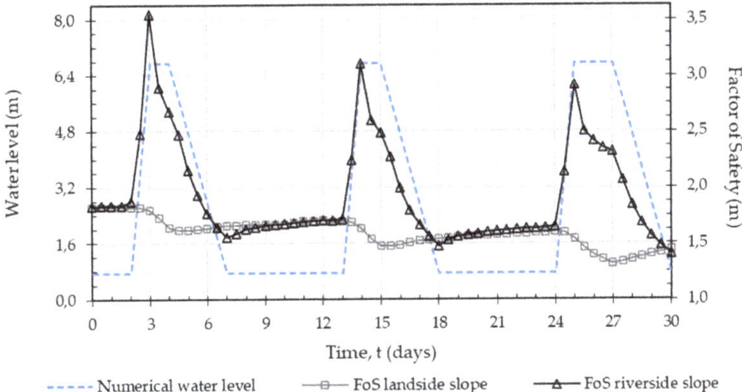

Figure 17. Variation over time of Factor of Safety computed at each time step for the critical slip surface of riverside and landside slopes, compared to the numerical water level assumed.

Following the first river stage rise, a reduction in the shear strength of the riverbank material is induced principally by the increasing of positive pore water pressures and the decrement in the resistance contribution related to soil suction. Hence, the FoS associated to the landside slope gradually decreases during the succession of hydrometric peaks, reaching a minimum value at last step of analysis.

With reference to the riverside slope, the stabilizing effect provided by the river water level prevails over the destabilizing forces generated by the build-up of pore water pressures when the water stage rises and, consequently, the FoS tends to increase. A significant decrease of the safety factor is however observed during the persistence of the hydrometric peaks, as a consequence of increasingly higher pore water pressures near the riverside slope. Then, when a rapid drawdown occurs, following a high water level stage, the embankment material is still close to a saturated condition, but the confining pressure is no longer present to counterbalance the shear strength reduction and this results in a fast decay of the safety factor.

Slip surfaces obtained by Limit Equilibrium stability analyses for both river embankment's sides are reported in Figure 18, through a graphical representation called 'safety map' [45]. Considering all the valid trial slip surfaces determined by the numerical code, these can be grouped into homogeneous ranges of FoS identified by different colored bands in order to easily identify critical zones. The red color is here associated to the lowest factors of safety, while the green represents the sliding mechanisms related to the highest FoS values.

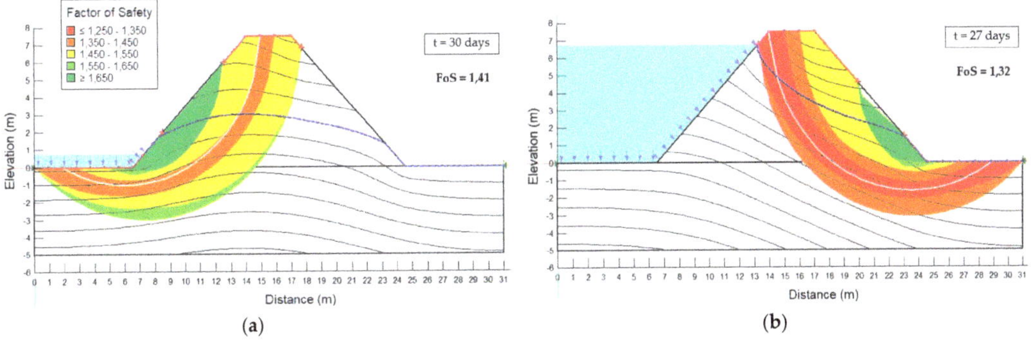

Figure 18. Riverbank safety maps for riverside (**a**) and landside (**b**) slopes, isolines of pore water pressure distribution for the most critical time steps and minimum FoS attained. The most critical slip surface is drawn in white.

Since the most critical slip surfaces for both riverside and landside slopes involve the upper portion of the foundation layer, starting from the crest of the riverbank, the predominant failure mechanism for the present embankment section might be the macro-instability of the slopes.

However, as the minimum value of the safety factor is approximately equal to 1.32 for the landside slope and to 1.41 for the riverside slope, the embankment results to be in a stable condition for the whole elapsed period of time. Although both values are far from unity, the riverside zone appears quite more secure than the landside one, as suggested by the color maps. Moreover, it is noted that the minimum value of safety factors is not attained at the same time in both slopes. It corresponds to the third hydrometric peak for the landside slope and to the third short drawdown for the riverside slope, as expected.

Comparing the results in terms of factor of safety between the steady-state and the transient seepage analyses carried out, it is evident that the steady-state condition is associated to a safety margin even lower, frequently resulting in a highly conservative river embankment stability assessment and, generally, as an excessively prudential assumption.

5. Conclusions

In the present contribution, a methodological approach for the investigation of the hydraulic response and the stability assessment of fluvial embankments, based on laboratory testing and numerical modelling has been thoroughly described. The entire procedure takes its beginnings from the synthetic representation of the main characteristics of riverbanks subjected to frequent fluctuations of retained water levels, where consequent variation in the positive pore water pressure and suction distribution are typical. A simplified geometry, eventually reproducible in scaled physical models and characterized by a trapezoidal embankment lying on a finer grained soil foundation, has been so taken as reference.

The laboratory setup starts with the identification of the optimal physical properties and the retention behaviour of the unsaturated compacted earthfill material and includes the hydraulic characterization of both layers involved in the study, as well as a series of triaxial compression tests in saturated conditions. Seepage and stability numerical simulations have been carried out through a combination of FE (for pore water pressure computation) and LE (for stability evaluation) analyses, firstly considering both the stationary and transient flow conditions and then focusing on two main global failure mechanisms of the riverside and landside slopes. The importance of carrying out transient seepage analyses, in the case of succeeding river flood events, has been highlighted, since steady-state seepage analyses performed at the maximum hydrometric height recorded, frequently results in excessively conservative stability assessments. Factor of safety variations with time have been determined, with respect to all these aspects, gathering key information on the effects of river level fluctuations on the embankment stability conditions, which can turn out very useful for further detailed experimental and numerical investigations on the river embankment response.

Author Contributions: Conceptualization, R.V. and M.P.; Data curation, R.V., E.D., C.G.G. and D.G.; Formal analysis, R.V. and E.D.; Investigation, R.V.; Methodology, R.V., E.D., C.G.G. and M.P.; Project administration, R.V.; Software, E.D., C.G.G. and M.P.; Supervision, R.V. and M.P.; Validation, R.V. and M.P.; Writing—original draft, R.V., E.D. and M.P.; Writing—review & editing, R.V., E.D., C.G.G., D.G. and M.P. All authors have read and agreed to the published version of the manuscript.

Funding: This research was funded under the scheme for "Research Projects of National Relevance" (in Italian: Progetti di Ricerca di Rilevante Interesse Nazionale—PRIN), Bando 2017, grant number 2017YPMBWJt, promoted by the Italian Ministry of Education, University and Research (in Italian: Ministero dell'Istruzione, dell'Università e della Ricerca—MIUR).

Institutional Review Board Statement: Not applicable.

Informed Consent Statement: Not applicable.

Data Availability Statement: The authors confirm that all data used during the study appear in the submitted article.

Acknowledgments: This work has been supported by the project "REDREEF-Risk Assessment of Earth Dams and River Embankment to Earthquakes and Floods" funded by the MIUR Progetti di Ricerca di Rilevante Interesse Nazionale (PRIN) Bando 2017-grant 2017YPMBWJt.

Conflicts of Interest: The authors declare no conflict of interest.

References

1. European Environment Agency. Floods. Available online: https://www.eea.europa.eu/archived/archived-content-water-topic/water-resources/floods (accessed on 23 April 2021).
2. European Environment Agency. Economic Losses from Climate-Related Extremes in Europe. Available online: https://www.eea.europa.eu/data-and-maps/indicators/direct-losses-from-weather-disasters-4/assessment (accessed on 23 April 2021).
3. Simonini, P.; Cola, S.; Bersan, S. Caratterizzazione Geotecnica, meccanismi di collasso e monitoraggio degli argini fluviali. In Proceedings of the XXV Convegno Nazionale di Geotecnica, Baveno, Italy, 4–6 June 2014; Volume 1, pp. 229–268.
4. CIRIA; Ministry of Ecology of United Kingdom; USACE. *The International Levee Handbook (C731)*; CIRIA: London, UK, 2013.
5. Vannucchi, G.; Gottardi, G.; Madiai, C.; Marchi, M.; Tonni, L. Analisi della probabilità di collasso arginale dei grandi fiumi. In Proceedings of the XXV Convegno Nazionale di Geotecnica, Baveno, Italy, 4–6 June 2014; Volume 1, pp. 303–319.
6. Wolman, M.G. Factors influencing the erosion of cohesive river banks. *Am. J. Sci.* **1959**, *257*, 204–216. [CrossRef]
7. Hooke, J.M. An analysis of the processes of river bank erosion. *J. Hydrol.* **1979**, *42*, 39–62. [CrossRef]
8. Thorne, C. Processes and mechanisms of river bank erosion. In *Gravel-Bed Rivers: Fluvial Processes, Engineering and Management*; John Wiley & Sons Inc.: Hoboken, NJ, USA, 1982; pp. 227–271.
9. Rinaldi, M.; Casagli, N. Stability of streambanks formed in partially saturated soils and effects of negative pore water pressures: The Sieve River (Italy). *Geomorphology* **1999**, *26*, 253–277. [CrossRef]
10. Federal Emergency Management Agency. *Technical Manual for Dam Owners: Impacts of Animals on Earthen Dams*; FEMA Rep. 47; FEMA: Washington, DC, USA, 2005.
11. Kok, M.; TU Delft/HKV lijn in water; Jongejan, R.; Jongejan, R.M.C.; Nieuwjaar, M.; Waternet; Tánczos, I. Ministerie van Infrastructuur en Milieu, Rijkswaterstaat, Expertise Netwerk Waterveiligheid (RWS, ENW). *Fundamentals of Flood Protection*; NPN Drukkers: Breda, The Netherlands, 2017.
12. Lee, C.W.; Kim, Y.S.; Park, S.Y.; Kim, D.G.; Heo, G. The behavior characteristics of a reservoir levee subjected to increasing water levels. *J. Civ. Eng. Manag.* **2017**, *23*, 15–27. [CrossRef]
13. Saran, R.K.; Viswanadham, B.V.S. Centrifuge model tests on the use of geosynthetic layer as an internal drain in levees. *Geotext. Geomembr.* **2018**, *46*, 257–276. [CrossRef]
14. Wang, Z.F.; Li, J.H.; Zhang, L.M. Influence of cracks on the stability of a cracked soil slope. In Proceedings of the 5th Asia-Pacific Conference on Unsaturated Soils, Pattaya, Thailand, 29 February–2 March 2012; Volume 2, pp. 594–600.
15. Orlandini, S.; Moretti, G.; Albertson, J.D. Evidence of an emerging levee failure mechanism causing disastrous floods in Italy. *Water Resour. Res.* **2015**, *51*, 7995–8011. [CrossRef]
16. Springer, F.M., Jr.; Ullrich, C.R.; Hagerty, D.J. Streambank stability. *J. Geotech. Eng.* **1985**, *111*, 624–640. [CrossRef]
17. Lawler, D.M.; Thorne, C.R.; Hooke, J.M. Bank erosion and instability. In *Fluvial Geomorphology for River Engineering and Management*; Thorne, C.R., Hey, R.D., Newson, M.D., Eds.; John Wiley & Sons: Chichester, UK, 1997; pp. 137–172.
18. Baldi, G.; Bellotti, R.; Crippa, V.; Fretti, C.; Ghionna, V.N.; Jamiolkowski, M.; Ostricati, D.; Pasqualini, E.; Pedroni, S. Laboratory validation of in situ tests. In Proceedings of the 11th International Conference on Soil Mechanics and Foundation Engineering, San Francisco, CA, USA, 12–16 August 1985.
19. Baldi, G.; Bellotti, R.; Ghionna, V.N.; Jamiolkowski, M.; Pasqualini, E. Interpretation of CPT's and CPTU's. In Proceedings of the 4th International Geotechnical Seminar, Balkema, Singapore, 27 November 1986; pp. 3–10.
20. Bellotti, R.; Jamiolkowski, M.; Lo Presti, D.C.F.; O'Neill, D.A. Anisotropy of small strain stiffness in Ticino sand. *Géotechnique* **1996**, *46*, 115–131. [CrossRef]
21. Fioravante, V. Anisotropy of small strain stiffness of Ticino and Kenya sands from seismic wave propagation measured in triaxial testing. *Soils Found.* **2000**, *40*, 129–142. [CrossRef]
22. Fioravante, V.; Giretti, D. Unidirectional cyclic resistance of Ticino and Toyoura sands from centrifuge cone penetration tests. *Acta Geotech.* **2016**, *11*, 953–968. [CrossRef]
23. Jamiolkowski, M.B.; Lo Presti, D.C.F.; Manassero, M. Evaluation of relative density and shear strength from CPT and DMT. In Proceedings of the Soil behavior and soft ground construction, Ladd symposium, Reston, VA, USA, 5–6 October 2001; Germaine, T.C., Sheahan, T.C., Whitman, R.V., Eds.; ASCE: Reston, VA, USA, 2001; Volume 119, pp. 201–238.
24. ASTM D1557-12e1. *Standard Test Methods for Laboratory Compaction Characteristics of Soil Using Modified Effort (56,000 ft-lbf/ft3 (2700 kN-m/m3))*; ASTM International: West Conshohocken, PA, USA, 2012; Available online: http://www.astm.org (accessed on 23 April 2021).
25. Darcy, H. *Les Fontaines Publiques de la Ville de Dijon: Exposition et Application des Principes a Suivre et des Formules a Employer dans les Questions de Distribution d'eau*; Dalmont: Paris, France, 1856.

26. Nicotera, M.; Papa, R.; Urciuoli, G. An experimental technique for determining the hydraulic properties of unsaturated pyroclastic soils. *Geotech. Test. J.* **2010**, *33*. [CrossRef]
27. Schindler, U. Ein schnellverfahren zur messung der wasserleitfähigkeit im teilgesättigten boden an stechz, Arch. *Acker Pfla-zenbau Bodenkd* **1980**, *24*, 1–7.
28. Richards, L.A. Porous plate apparatus for measuring moisture retention and transmission by soils. *Soil Sci.* **1948**, *66*, 105–110. [CrossRef]
29. Richards, L.A. Physical condition of water in soil. In *Methods of Soil Analysis. Part 1. Agronomy Monograph*; Black, C.A., Ed.; ASA, SSSA: Madison, WI, USA, 1965; pp. 128–152.
30. Mualem, Y. A new model for predicting the hydraulic conductivity of unsaturated porous media. *Water Resour. Res.* **1976**, *12*, 513–522. [CrossRef]
31. Van Genuchten, M.T. A closed-form equation for predicting the hydraulic conductivity of unsaturated soils. *Soil Sci. Soc. Am. J.* **1980**, *4*, 892–898. [CrossRef]
32. Šimůnek, J.; Šejna, M.; Saito, H.; Sakai, M.; van Genuchten, M.T. *HYDRUS Software Series 3*, Version 4.0; The HYDRUS-1D Software Package for Simulating the Movement of Water, Heat, and Multiple Solutes in Variably Saturated Media; Department of Environmental Sciences, University of California Riverside: Riverside, CA, USA, 2008.
33. Marquardt, D.W. An algorithm for least-squares estimation of nonlinear parameters. *J. Soc. Ind. Appl. Math.* **1963**, *11*, 431–441. [CrossRef]
34. Pirone, M.; Reder, A.; Rianna, G.; Pagano, L.; Nicotera, M.V.; Urciuoli, G. Laboratory and physical prototype tests for the investigation of hydraulic hysteresis of pyroclastic soils. *Geosciences* **2020**, *10*, 320. [CrossRef]
35. Rianna, G.; Comegna, L.; Pagano, L.; Picarelli, L.; Reder, A. The role of hydraulic hysteresis on the hydrological response of pyroclastic silty covers. *Water* **2019**, *11*, 628. [CrossRef]
36. Hueckel, T.; Baldi, G. Thermoplasticity of saturated clays: Experimental constitutive study. *J. Geotech. Eng.* **1990**, *116*. [CrossRef]
37. Hueckel, T.; Pellegrini, R. Thermoplastic modelling of undrained failure of saturated clay due to heating. *Soils Found.* **1991**, *31*, 11–16. [CrossRef]
38. Geo-slope International Ltd. *Seep/W, Groundwater Seepage Analysis*; GEO-SLOPE International Ltd.: Calgary, AB, Canada, 2012.
39. Fredlund, D.G.; Rahardjo, H. *Soil Mechanics for Unsaturated Soils*; John Wiley & Sons: New York, NY, USA, 1993.
40. Richards, L.A. Capillary conduction of liquids through porous medium. *Int. J. Appl. Phys.* **1931**, *1*, 318–333. [CrossRef]
41. Gottardi, G.; Gragnano, C.G.; Ranalli, M.; Tonni, L. Reliability analysis of riverbank stability accounting for the intrinsic variability of unsaturated soil parameters. *Struct. Saf.* **2020**, *86*, 01973. [CrossRef]
42. Geo-slope International Ltd. *Slope/W, Slope Stability Analysis*; GEOSLOPE International Ltd.: Calgary, AB, Canada, 2012.
43. Morgenstern, N.R.; Price, V.E. The analysis of the stability of general slip surfaces. *Géotechnique* **1965**, *15*, 79–93. [CrossRef]
44. Vanapalli, S.K.; Fredlund, D.G.; Pufahl, D.E.; Clifton, A.W. Model for the prediction of shear strength with respect to soil suction. *Can. Geotech. J.* **1996**, *33*, 379–392. [CrossRef]
45. Baker, R.; Leshchinsky, D. Spatial distribution of safety factors. *J. Geotech. Geoenviron.* **2001**, *127*, 135–145. [CrossRef]

geoscienes

Article

Capillary Barriers during Rainfall Events in Pyroclastic Deposits of the Vesuvian Area

Ciro Sepe [1,2], Domenico Calcaterra [1], Manuela Cecconi [3,*], Diego Di Martire [1], Lucia Pappalardo [4], Riccardo Scarfone [5,6], Enza Vitale [1] and Giacomo Russo [1]

1   Department of Earth Science, Environment and Resources, University of Napoli Federico II, 80126 Napoli, Italy; ciro.sepe@unina.it (C.S.); domcalca@unina.it (D.C.); diego.dimartire@unina.it (D.D.M.); enza.vitale@unina.it (E.V.); giarusso@unina.it (G.R.)
2   Department of Civil and Mechanical Engineering, University of Cassino and Southern Lazio, 03043 Cassino, Italy
3   Department of Engineering, University of Perugia, 06125 Perugia, Italy
4   INGV/Osservatorio Vesuviano—Napoli, 80125 Napoli, Italy; lucia.pappalardo@ingv.it
5   Geotechnical Consulting Group, London SW7 5BE, UK; riccardo.scarfone@glasgow.ac.uk
6   James Watt School of Engineering, University of Glasgow, Glasgow G12 8QQ, UK
*   Correspondence: manuela.cecconi@unipg.it

check for
updates

Citation: Sepe, C.; Calcaterra, D.; Cecconi, M.; Di Martire, D.; Pappalardo, L.; Scarfone, R.; Vitale, E.; Russo, G. Capillary Barriers during Rainfall Events in Pyroclastic Deposits of the Vesuvian Area. *Geosciences* **2021**, *11*, 274. https://doi.org/10.3390/geosciences11070274

Academic Editors: Roberto Vassallo, Luca Comegna, Roberto Valentino and Jesus Martinez-Frias

Received: 12 May 2021
Accepted: 23 June 2021
Published: 29 June 2021

Publisher's Note: MDPI stays neutral with regard to jurisdictional claims in published maps and institutional affiliations.

Abstract: In the present paper, the capillary barrier formation at the interface between soil layers, which is characterized by textural discontinuities, has been analyzed. This mechanism has been investigated by means of a finite element model of a two-layer soil stratification. The two considered formations, belonging to the pyroclastic succession of the "Pomici di Base" Plinian eruption (22 ka, Santacroce et al., 2008) of the Somma–Vesuvius volcano, are affected by shallow instability phenomena likely caused by progressive saturation during the rainfall events. This mechanism could be compatible with the formation of capillary barriers at the interface between layers of different grain size distributions during infiltration. One-dimensional infiltration into the stratified soil was parametrically simulated considering rainfall events of increasing intensity and duration. The variations in the suction and degree of saturation over time allowed for the evaluation of stability variations in the layers, which were assumed as part of stratified unsaturated infinite slopes.

Keywords: pyroclastic soils; infiltration; capillary barriers; stability analysis

1. Introduction

Infiltration processes through layered soil deposits can be deeply affected by the presence of layers with contrasting hydraulic properties. In saturated conditions, the hydraulic conductivity of an upper, finer-grained layer overlying a coarser-grained layer can be higher than that of the coarser layer; however, in unsaturated conditions, the situation can be the opposite. The coarser layer can hinder the infiltration process through reduced unsaturated hydraulic conductivity, thus promoting the formation of a capillary barrier. This principle is applied in artificial layered soil covers to develop capillary barriers that prevent infiltration and promote lateral drainage (e.g., for landfill covers [1] and slope stabilization [2]).

Layered pyroclastic soil profiles are often complex, owing to the different eruptive phases and the deposition process of volcanic materials that follows. This is the case for the pyroclastic materials spread out over a large part of the mountainous areas surrounding the volcanic centers of Campania (southern Italy). Soil covers are indeed characterized by alternating layers of ashes, pumices, scoriae, and other volcanic products, which presents significant differences in terms of the geochemical nature, texture, density, hydraulic and mechanical properties [3,4].newpage Pyroclastic soil covers of Campania are frequently involved in rainfall-induced shallow landslides. Multiple factors can trigger landslides among these materials, including the site conditions (geomorphological characteristics, rainfall patterns, etc.) and properties (e.g., porosity, particle size distributions) of the pyroclastic soils [5–8].

The spatial variations in the textural characteristics and the related hydro-mechanical properties of the pyroclastic layers are the key factors controlling the rates of water infiltration, with the potential to form capillary barriers at the interface between layers of different soil textures (e.g., [1,9]). Capillary barriers have been considered by some authors to be one of the factors influencing the possible instability of shallow landslides [10,11]. Several studies [10–14] have shown that the differences in texture between the layers of the pyroclastic covers of Campania may affect infiltration processes and, eventually, slope instability.

The in situ observation of capillary barrier formation is a difficult task, due to the transient nature of the phenomenon and the fact that it is highly dependent on the initial soil-moisture profiles and boundary conditions, as shown by several laboratory experiments in soil columns, lysimeters, and instrumented flumes [15–19]. Numerical simulations represent an alternative to experimental measurements when examining the performance of the capillary barrier system. Oldenburg and Pruess [20] comprehensively studied the behavior of capillary barriers through the use of numerical methods. Stormont and Morris [21] and Khire et al. [1] performed parametric numerical analyses on horizontal capillary systems to investigate the effects of different parameters on the water storage capacity and hydraulic response of capillary barrier systems when subjected to different environmental loads. Numerical analyses were conducted by Rahardjo et al. [22] to determine the thickness and length of the fine- and coarse-grained materials of the capillary barrier system. Scarfone et al. [23,24] used advanced numerical modeling to extensively study the effectiveness of capillary barrier systems in introducing water retention hysteresis into the hydraulic behavior of unsaturated soils, considering a new hydraulic conductivity model at a low degree of saturation, and the complex phenomenon of soil–atmosphere interaction in the long term.

In the present paper, the capillary barrier formation at the interface between soil layers, which is characterized by textural discontinuities, has been investigated by means of a finite element model. Two layers of soil stratification belonging to the pyroclastic succession of the "Pomici di Base" Plinian eruption (22 ka, Santacroce et al. [25]) of the Somma–Vesuvius volcano have been considered in the numerical model. The stratification is affected by shallow instability phenomena [26] that are likely caused by the progressive saturation during the rainfall events; this is a mechanism that is compatible with the formation of capillary barriers at the interface between layers of contrasting hydraulic properties during infiltration. The water retention properties and hydraulic conductivity function were derived for the two layers, and one-dimensional infiltration was parametrically simulated considering rainfall events of increasing intensity and duration. Variations in the suction and degree of saturation over time allowed for the evaluation of the stability of the stratification assumed as a stratified unsaturated infinite slope.

2. Pyroclastic Soils

The territory around the town of Palma Campania, about 20 km east of the city of Naples, Southern Italy, is located in the "Conca Napoletana" (Campanian plain) at the border of the Nola and Sarno plains; it is delimited to the East by the Apennine Mesozoic carbonate reliefs and to the West by the Somma–Vesuvius Quaternary volcanic complex (Figure 1).

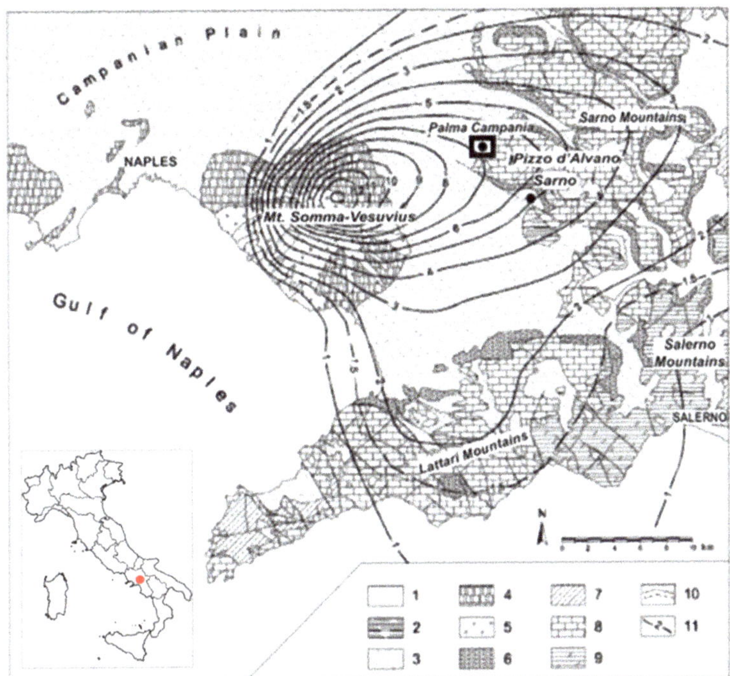

Figure 1. Geological map of the southwestern portion of the Campanian plain: (1) alluvial deposits, (2) travertine, (3) loose ash-fall deposits, (4) mainly coherent ash-flow deposits, (5) lavas, (6) debris and slope talus deposits, (7) Miocene flysch, (8) middle Jurassic–Upper Cretaceous limestones, (9) lower Triassic–Middle Jurassic dolomites and calcareous limestones, (10) fault, (11) total isopach lines (m) of the most important Mt. Somma–Vesuvius explosive eruptions. This figure is modified from [5].

On the carbonate slopes surrounding Palma Campania, relatively recent small landslides of the translational sliding type [27] have been observed among pyroclastic materials. A detailed field survey allowed for the identification of a series of recent landslides, mainly located in the "Vallone Lupici", involving the Somma–Vesuvius pyroclastic deposits. The failure surface has been identified at the interface between the grey pumice and dark scoria layers (Figure 2). This is possibly due, in large part, to the substantial variation in the grain size distributions and hydraulic properties between these two levels, such that water infiltration is affected.

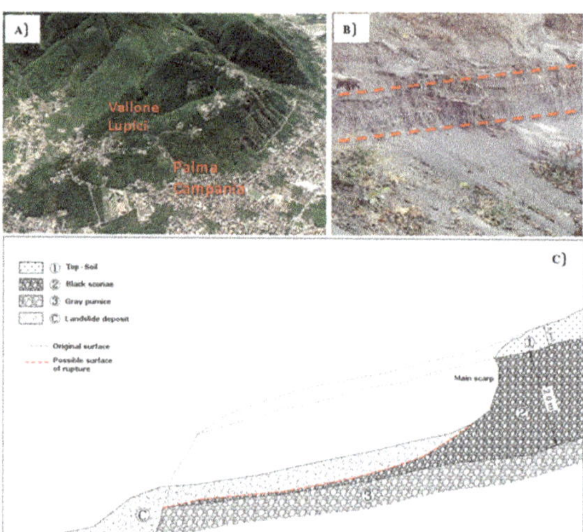

Figure 2. (**A**) A 3D satellite image of the studied area (from Google Earth); (**B**) the detachment zone of a landslide in Vallone Lupici; and (**C**) the cross-section of a typical landslide.

The studied volcanic succession (Figure 3) includes a few-meters-thick fallout deposit consisting—from the base to the top—of a basal, high-vesiculated, white pumice layer; an intermediate level of high to moderate vesiculated grey pumices; and an upper level of denser, black scoriae. This stratigraphic sequence corresponds to fallout deposits emplaced during the so-called "Pomici di Base" Plinian eruption (22,000 yrs BP [25]) of the Somma–Vesuvius volcano, which is well described in the literature [28–31].

Figure 3. Vallone Lupici pyroclastic deposits—the identification of the Pomici di Base Plinian Eruption layers: BS, black scoriae; GPs, grey pumices; WPs, white pumices.

The observed variations in lithological facies in the stratigraphy of this fallout deposit well-match the changes in eruption dynamics during the progression of the volcanic activity. In fact, the reduction in the clast vesicularity and size at the top of the sequence reflects the existence of a compositional gradient in the magma chamber. Whereas the first explosive phases tap the volatile-rich top of the felsic magma body (forming the first deposited materials at the base of the sequence), the following less-explosive phases sample partially degassed mafic magmas at the bottom of the reservoir (deposited later in the upper stratigraphic levels).

Bulk samples were collected at different stratigraphic heights, consistent with variations in lithological facies (i.e., grain size, color, and texture). Moreover, relatively undisturbed whole samples were retrieved using thin-walled tube samplers, from the zone close to the failure surface, specifically at the transition between the grey pumices and the black scoriae layer.

Grain size distributions of samples vary along the stratigraphic sequence. As reported in Figure 4—according to ASTM D422—black scoriae consist of 80% sand and 20% gravel, grey pumices consist of 80% gravel and 20% sand, and white pumices consist of 70% gravel and 30% sand.

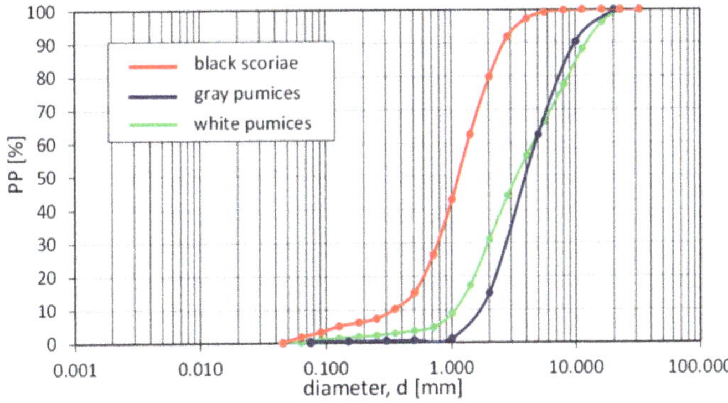

Figure 4. Grain size distributions of the investigated units in stratigraphic succession.

The specific weight of solid grains (γ_s) is 27.8 kN/m^3 for black scoriae, 27.2 kN/m^3 for grey pumices, and 24 kN/m^3 for white pumices. Variations in the density of the solid phase are consistent with the geochemical changes described in the literature [29], from the dense trachybasaltic-latitic scoriae of the top layer to the light trachytic pumices at the bottom of the volcanic succession.

Bulk weight measurements, carried out on at least 100 single clasts of each representative class (namely 2 mm, 5 mm, and 10 mm), showed no significant variations within each stratigraphic unit [26]. The dry specific weight (γ_d) increased from the base to the top of the stratigraphic succession, with lower average values pertaining to basal white pumices ($\gamma_d = 6$ kN/m^3), intermediate values pertaining to grey pumices ($\gamma_d = 12.4$ kN/m^3), and larger values pertaining to black scoriae ($\gamma_d = 15$ kN/m^3). Intergranular porosities, determined using a microstructural analysis performed by means of X-ray tomography on the reconstituted sample [26], were estimated to be equal to 33.8% for black scoriae and 36.5% for grey pumices.

3. Water Retention Properties

The water retention properties of the pyroclastic soil under study were determined using the model proposed by Arya and Paris [32], based on the particle size distribution of each formation in the stratigraphic sequence. In the model proposed by Arya and Paris, it is assumed that the solid grains are spherical, and the pore volume is approximated to that of cylindrical capillary tubes. For each ith particle-size class, the pore radius (r_i)—associated with the mean grain radius (R_i)—is related to the suction value s_i according to:

$$r_i = R_i \left[2en_i^{(1-\alpha)}/3 \right]^{1/2} \text{ and } s_i = \frac{2T_w}{r_i} \tag{1}$$

where n_i, e, T_w, and α are, respectively, the number of spherical grains, the void ratio, the surface tension of water ($T_w = 7.27 \times 10^{-2}$ N/m at 20 °C), and a constant parameter larger than unity. Thus, for a given grain size distribution, Equation (1) allows for the calculation of the value of the suction required to desaturate a given fraction of pores. In the following, α is assumed to be constant, as consistent with the one evaluated for other typical pyroclastic weak rocks and the coarse-grained soils of Central Italy [33,34].

The Arya and Paris model was applied to two layers of the considered stratigraphic succession, namely to the black scoriae (BS) and grey pumices (GP), which form the interface of interest for the possible formation of a capillary barrier during an infiltration process. From the application of the method described above, volumetric water content values were obtained as a function of suction, which represents the behavior of the materials during wetting. These values were expressed in terms of the effective degree of saturation (S_e), assuming a residual degree of saturation for both soils equal to zero, and were interpolated using the function proposed by Van Genuchten [35]:

$$S_e = \left[\frac{1}{1 + (as)^n} \right]^m \tag{2}$$

where a, n, and m are fitting parameters, reported in Table 1 for both GP and BS, along with fitting parameters proposed in the literature for similar coarse layers. Soil water retention curves for both the BS and GP layers are reported in Figure 5.

Table 1. Fitting parameters of the Van Genuchten model.

	This Study		Damiano et al. (2012)	De Vita et al. (2013)	Mancarella et al. (2012)
	BS	**GP**			
a [kPa^{-1}]	2.2	18.1	3.3	4.2	28
n	1.8	2.0	7	1.43	2.5
m	0.5	0.5	0.5	0.3	0.6
S_r	0	0	0.5	0	0

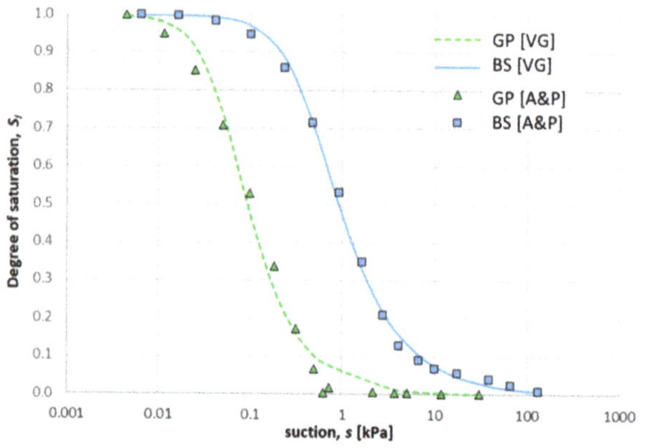

Figure 5. Water retention curves of GP and BS soils, estimated using the Arya and Paris (AP) model and the Van Genuchten (VG) model.

4. Infiltration Process

The implemented numerical model, aimed at simulating the formation of a capillary barrier at the interface between two layers, is characterised by textural discontinuities. A parametric study of infiltration was carried out by varying the intensity and duration of the considered rainfall events.

The hydraulic constitutive Van Genuchten model for unsaturated soils was implemented in the CODE_BRIGHT finite-element code [36,37]. The numerical model was a vertical column of soil made of two layers: an upper layer, 2.5 m thick, representing the finer layer (BS) of a capillary barrier system, and a lower layer, 0.5 m thick, representing the coarser layer (GP) (Figure 6). The materials forming the two layers were each modeled by defining the soil water retention curve and the porosity. The intergranular porosities of reconstituted samples of BS and GP were assumed to be representative for each layer. The potential hydraulic effects of double porosity (i.e., intergranular and intragranular porosity) were not considered. Each of the two layers was treated as a uniform material. The hydraulic conductivity function was derived from the Mualem model [38].

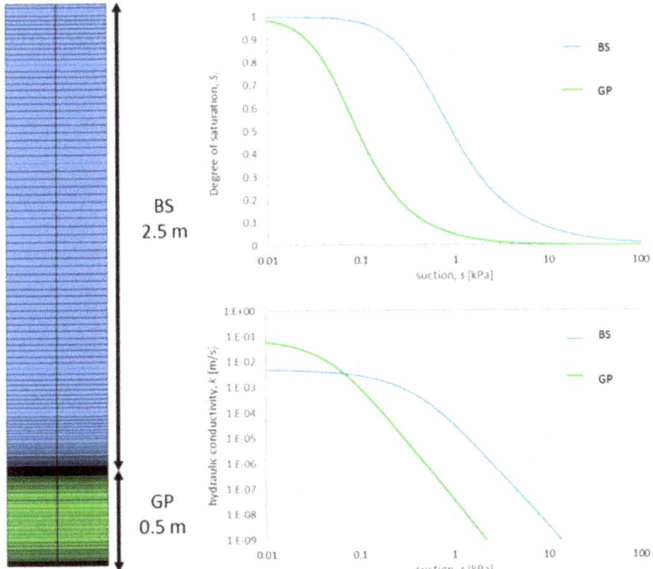

Figure 6. The geometry of the two-layer numerical model, Van Genuchten water retention curves, and Mualem hydraulic conductivity.

During numerical simulations of one-dimensional infiltration, only isothermal liquid transport was considered, with the solid phase treated as nondeformable and the gas phase as nonmobile. Thus, constant and uniform values of temperature (T = 20 °C), displacements of the solid phase (u = 0 m), and gas pressure (u_a = 0 kPa) were imposed.

The initial condition for the numerical analyses was a hydrostatic suction profile with s = 20 kPa at the bottom boundary, s = 50 kPa at the top, and a linear variation between these values, with null water pressure at depth equal to 2.0 m below the bottom boundary (z_w = 5 m from the ground table). In this initial condition, the coarser layer was at a very low degree of saturation. For the bottom boundary condition, a free drainage condition was imposed. This consists of a zero water flow value if the suction is higher than 0.3 kPa, whereas a suction value of 0.3 kPa is imposed if the suction at the bottom boundary attains this value. The value of 0.3 kPa is the suction value at which the lower coarser layer GP starts draining water (see hydraulic properties in Figure 6). For the top boundary condition, a constant value of vertical water flux (the infiltration rate) was imposed. To assess the influence of the infiltration rate, different rainfall scenarios—in terms of rainfall intensity (5 mm/h and 10 mm/h) and duration (12 h and 24 h)—were considered. The selected rainfall events have been considered to be representative events at the site, as highlighted by the data shown in Figure 7, which was acquired across approximately two years of monitoring.

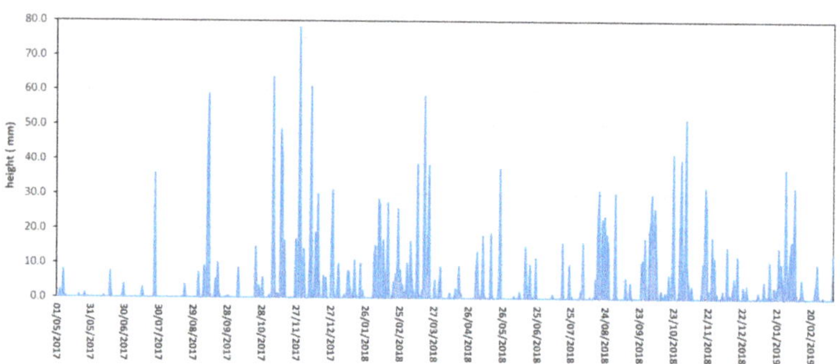

Figure 7. Rainfall events monitored at the site during the time interval from 05/2017–02/2019.

5. Results and Discussion

5.1. Infiltration Process

The infiltration analysis is carried out with reference to the suction and degree of saturation profiles, and by checking the water storage in the upper BS layer. These parameters are controlled over time, beginning at the start of the considered rainfall event.

For a rainfall event with an intensity of 5 mm/h and a duration of 12 h, it is observed that (Figure 8) the infiltration front proceeding downward causes a reduction in suction and an increase in the degree of saturation along the vertical column for periods of time that are longer than the duration of the rainfall event. When the interface is reached (depth of 2.5 m), the degree of saturation increases and reaches its maximum value in the upper layer, while it remains almost constant in the lower layer. Similarly, suction decreases at the interface in the BS layer and reaches the minimum value, while it remains at the initial value in the grey pumices. The conditions described correspond with the formation of a capillary barrier at the interface between black scoriae and grey pumices. The capillary barrier also persists for a longer time, as evidenced by both the suction and saturation degree profiles at observation time t = 480 h. Further confirmation of this permanence comes from the constant value over time of the water storage in the BS layer, as shown in Figure 9.

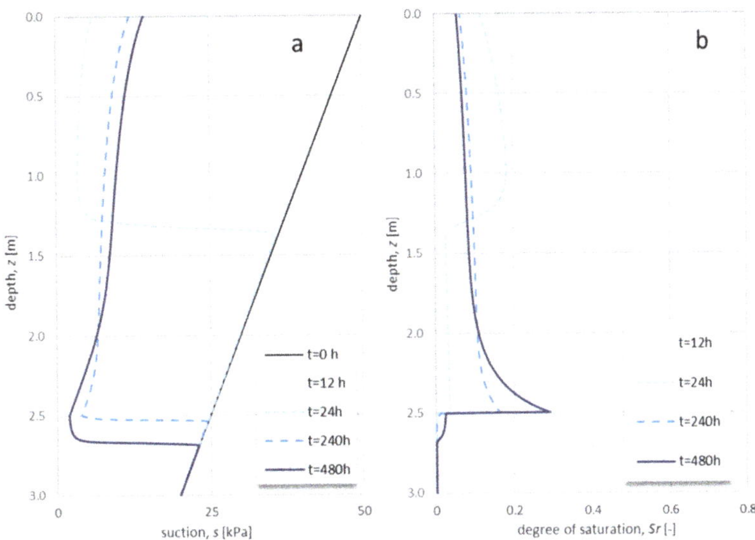

Figure 8. Infiltration process for a rainfall event with an intensity of 5 mm/h and a duration of 12 h: (**a**) the suction profile, and (**b**) the degree of saturation profile.

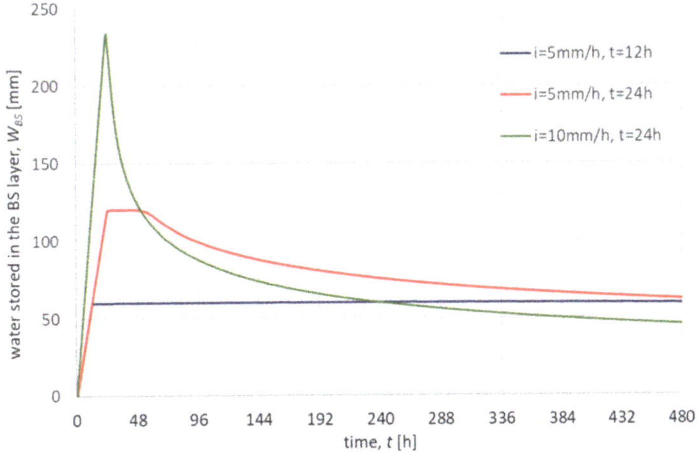

Figure 9. Water stored in the BS layer during the simulated infiltration processes.

For a rainfall event with an intensity of 5 mm/h and a duration of 24 h, the infiltration front close to the interface causes an increase in the degree of saturation (Figure 10a); however, this is coupled with suction reduction at the interface. The suction attained at the interface is so low that break-through has occurred in the capillary barrier, and water flow takes place toward the lower formation, with a progressive reduction in water storage in the upper black scoriae layer, as highlighted in Figure 9. As reported in Figure 11, break-through occurs when suction at the interface drops down from 25 kPa to 0.6 kPa, which is close to the theoretical value (s = 0.46 kPa) proposed by Lu and Likos [39].

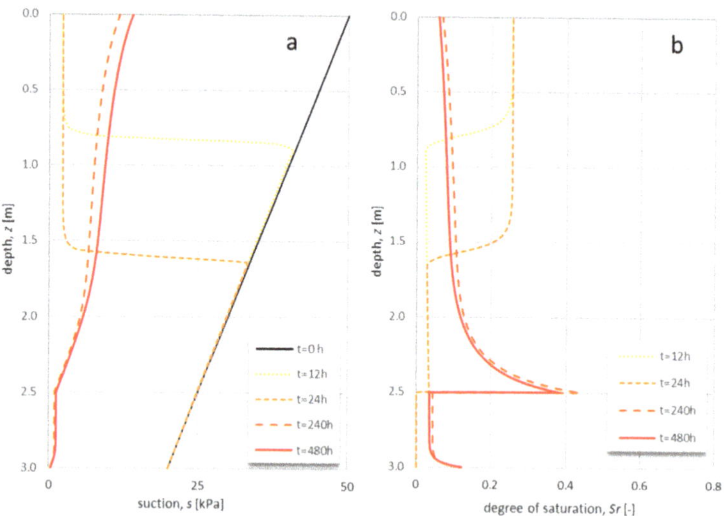

Figure 10. Infiltration process for a rainfall event with an intensity of 5 mm/h and a duration of 24 h: (**a**) the suction profile, and (**b**) the degree of saturation profile.

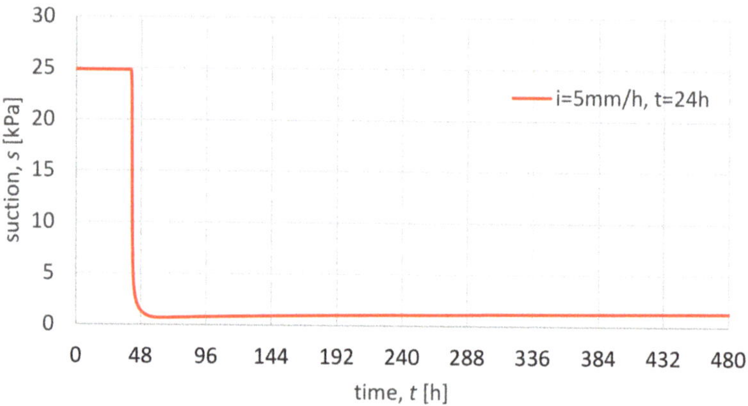

Figure 11. Suction evolution at the interface between BS and GP during the rainfall event at an intensity of 5 mm/h and a duration of 24 h.

For a rainfall event with an intensity of 10 mm/h and a duration of 24 h (Figure 12), the infiltration front crosses the interface in less time than the duration of the rainfall event. The observed increase in the degree of saturation in the BS layer approaching the interface is due to the contrast in the hydraulic conductivities of the two layers. The reduction in water storage over time, beginning at the end of the rainfall event, is a further confirmation that the capillary barrier effect at the interface is rapidly lost for the simulated rainfall event.

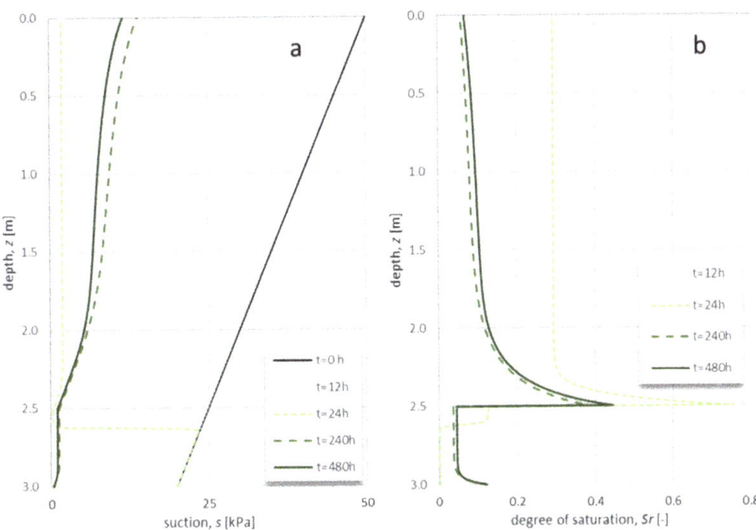

Figure 12. Infiltration process for a rainfall event with an intensity of 10 mm/h and a duration of 24 h: (**a**) the suction profile, and (**b**) the degree of saturation profile.

5.2. Stability Analyses

This section refers to the results obtained from a numerical example of application, which was aimed at highlighting the hydraulic effects on slope stability conditions following a rainfall event lasting 24 h and at an intensity equal to 5 mm/h and 10 mm/h, respectively. The case study is a slope (slope angle $\alpha = 36°$) of black scoriae layer ($\varphi = 35°$, $c' = 0$, $\gamma = 15$ kN/m^3) 2.5 m thick superposed to a bottom layer of grey pumices ($\varphi = 37°$, $c' = 0$, $\gamma = 12.4$ kN/m^3) 0.5 m thick. The shear strength parameters of the two layers have been deduced by [5] and [40]. The water table is located at a depth of $z_w = 5$ m from the ground table. The profiles of the saturation degree and soil suction during the infiltration events are shown in Figure 7, Figure 9, and Figure 10 at different times of observation ($t = 12$ h, 24 h, 240 h, and 480 h). These profiles, in turn, made it possible to estimate the soil shear strength at increasing depths, by adopting the failure criterion for unsaturated cohesionless soils proposed by Vanapalli et al. [41]. This leads to Equation (3), below:

$$\text{SF}(z) = \frac{\tau_f}{\tau} = \left(1 + \frac{S_r(z) \cdot s(z)}{\gamma z \cos^2 \alpha}\right) \cdot \frac{\tan \varphi}{\tan \alpha} \tag{3}$$

The decrease in the safety factor with depth, as the infiltration front proceeds, has been shown in Figure 13 as a consequence of the reduction in the positive contribution provided by the degree of saturation and suction in the expression (3). For a slope inclination close to the value of the internal friction angle of the shallow layer, it can be observed that for a rainfall event with an intensity (i) of 5 mm/h and a duration of 24 h, the factor of safety is progressively reduced as the infiltration front proceeds, but no instability of the slope occurs. At longer observation times, when the front reaches the interface between black scoriae and grey pumices, the formation of the capillary barrier (as described in Section 5.1 and shown in Figure 10) favors the further reduction in the safety factor—down to the instability condition—which is evident for the observation times of 240 h and 480 h. The failure surface is localized within the black scoriae layer, while the underlying grey pumices layer remains stable ($\text{SF}_{min} = 1.07$).

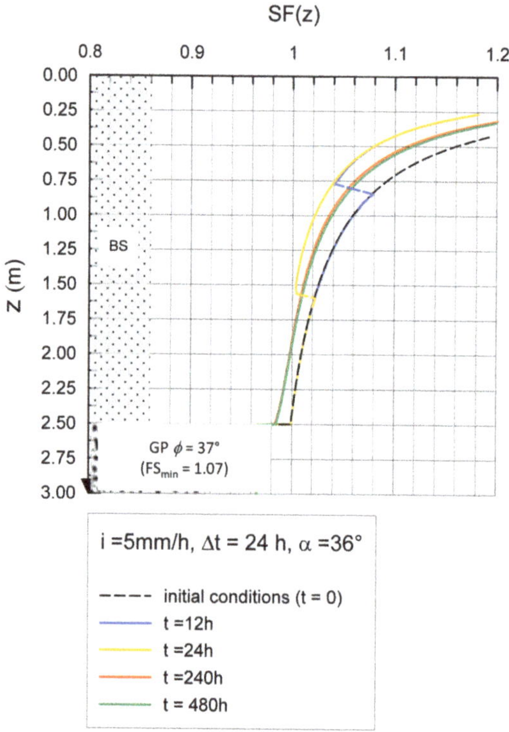

Figure 13. Safety factor (SF) vs. depth (z) for a rainfall event with an intensity (i) of 5 mm/h and a duration of 24 h, at increasing observation times.

For a rainfall event at a higher intensity (i = 10 mm/h) and a duration of 24 h, the infiltration front reaches the interface during the rainfall event, and the formation of the capillary barrier favors the onset of the unstable condition during the 24 h of a rainfall event.

A stability analysis of the infinite slope model (for stability conditions close to failure) showed how the formation and permanence of the capillary barrier can induce the instability of the shallow layer, confirming the in situ observation of the localization of instabilities in the shallow finer layers. The failure condition is triggered by rainfall events at intensity and duration levels favoring water storage in the upper layer and minimizing the beneficial contribution of suction to the stability of the slope.

6. Conclusions

A detailed field survey allowed for the identification of a series of recent landslides, mainly located in the "Conca Napoletana" (Campanian plain), involving the Somma–Vesuvius pyroclastic deposits. The failure surface has been identified at the interface between grey pumice and black scoriae layers, likely due to the contrasting hydraulic properties between these two levels, which promote the formation of capillary barriers. This hypothesis has been explored in the present paper by means of a numerical study of water infiltration, parametrically simulated considering rainfall events of increasing intensity and duration. The numerical analyses showed the effects of progressive water storage in the upper, finer-grained layer of black scoriae, which can give place to the capillary barrier at the interface with the coarser-grained layer of grey pumices. The time history of the infiltration process allowed the present analyses to highlight the capillary barrier water break-through as a consequence of progressive suction equalization between the two layers, as well as the gravity-driven downward flux of water over time. The impact of infiltration, and the influence of the capillary barrier on the layers' stability, has been evaluated with reference to the unsaturated infinite slope model. The attainment of failure conditions of the black scoriae layer for a medium intensity, long duration rainfall event is consistent with the field observations.

Author Contributions: Conceptualization and methodology, M.C., G.R., D.C., L.P. and D.D.M.; investigation, C.S.; data curation, C.S., R.S., L.P. and E.V.; writing—original draft preparation, G.R. and M.C.; writing—review and editing, D.C., D.D.M., L.P., R.S. and E.V.; supervision, M.C., G.R. All authors have read and agreed to the published version of the manuscript.

Funding: This research received no external funding.

Conflicts of Interest: The authors declare no conflict of interest.

References

1. Khire, M.V.; Benson, C.H.; Bosscher, P.J. Capillary barriers: Design variables and water balance. *J. Geotech. Geoenviron. Eng.* **2000**, *126*, 695–708. [CrossRef]
2. Rahardjo, H.; Satyanaga, A.; Leong, E.C. Unsaturated soil mechanics for slope stabilization. In Proceedings of the 5th Asia–Pacific Conference on Unsaturated Soils, Pattaya, Thailand, 29 February–2 March 2012; pp. 103–117.
3. Pappalardo, L.; Piochi, M.; Mastrolorenzo, G. The 3550 YR BP-1944 AD magma-plumbing system of Somma-Vesuvius: Constraints on its behavior and present state through a review of Sr-Nd isotope data. *Ann. Geophys.* **2004**, *47*, 1471–1483.
4. Pappalardo, L.; Mastrolorenzo, G. Short residence times for alkaline Vesuvius magmas in a multi-depth supply system: Evidence from geochemical and textural studies. *Earth Planet. Sci. Lett.* **2010**, *296*, 133–143. [CrossRef]
5. De Vita, P.; Napolitano, E.; Godt, J.W.; Baum, R.L. Deterministic estimation of hydrological thresholds for shallow landslide initiation and slope stability models: Case study from the Somma-Vesuvius area of southern Italy. *Landslides* **2013**, *10*, 713–728. [CrossRef]
6. Cascini, L.; Sorbino, G.; Cuomo, S.; Ferlisi, S. Seasonal effects of rainfall on the shallow pyroclastic deposits of the Campania region (southern Italy). *Landslides* **2014**, *11*, 779–792. [CrossRef]
7. Urciuoli, G.; Pirone, M.; Comegna, L.; Picarelli, L. Long-term investigations on the pore pressure regime in saturated and unsaturated sloping soils. *Eng. Geol.* **2016**, *212*, 98–119. [CrossRef]
8. Picarelli, L.; Olivares, L.; Damiano, E.; Darban, R.; Santo, A. The effects of extreme precipitations on landslide hazard in the pyroclastic deposits of Campania Region: A review. *Landslides* **2020**, *17*, 2343–2358. [CrossRef]
9. Shackelford, C.D.; Chang, C.-K.; Chiu, T.-F. The capillary barrier effect in unsaturated flow through soil barriers. In Proceedings of the First International Congress on Environmental Geotechnics, Edmonton, AB, Canada, 10–15 July 1994; pp. 789–793.
10. Mancarella, D.; Doglioni, A.; Simeone, V. On capillary barrier effects and debris slide triggering in unsaturated soil covers. *Eng. Geol.* **2012**, *147–148*, 14–27. [CrossRef]
11. Damiano, E. Effects of layering on triggering mechanisms of rainfall-induced landslides in unsaturated pyroclastic granular soils. *Can. Geotech. J.* **2018**, *56*, 1278–1290. [CrossRef]
12. Crosta, G.B.; Negro, P.D. Observations and modelling of soil slip-debris flow initiation processes in pyroclastic deposits: The Sarno 1998 event. *Nat. Hazards Earth Syst. Sci.* **2003**, *3*, 53–69. [CrossRef]
13. Vingiani, S.; Mele, G.; De Mascellis, R.; Terribile, F.; Basile, A. Volcanic soils and landslides: A case study of the island of Ischia (southern Italy) and its relationship with other Campania events. *Solid Earth* **2015**, *6*, 783–797. [CrossRef]
14. Reder, A.; Pagano, L.; Picarelli, L.; Rianna, G. The role of the lowermost boundary conditions in the hydrological response of shallow sloping covers. *Landslides* **2017**, *14*, 861–873. [CrossRef]
15. Yang, H.; Rahardjo, H.; Leong, E.C.; Fredlund, D.G. A study of infiltration on three sand capillary barriers. *Can. Geotech. J.* **2004**. [CrossRef]
16. Sharma, R.H.; Nakagawa, H. Numerical model and flume experiments of single- and two-layered hillslope flow related to slope failure. *Landslides* **2010**, *7*, 425–432. [CrossRef]
17. Mancarella, D.; Simeone, V. Capillary barrier effects in unsaturated layered soils, with special reference to the pyroclastic veneer of the Pizzo d'Alvano, Campania (Italy). *Bull. Eng. Geol. Environ.* **2011**, *71*, 791–801. [CrossRef]
18. Pagano, L.; Reder, A.; Rianna, G. Experiments to Investigate the Hydrological Behaviour of Volcanic Covers. *Procedia Earth Planet. Sci.* **2014**, *9*, 14–22. [CrossRef]
19. Tan, S.H.; Wong, S.W.; Chin, D.J.; Lee, M.L.; Ong, Y.H.; Chong, S.Y.; Kassim, A. Soil column infiltration tests on biomediated capillary barrier systems for mitigating rainfall-induced landslides. *Environ. Earth Sci.* **2018**, *77*, 589. [CrossRef]
20. Oldenburg, C.M.; Pruess, K. On Numerical Modeling of Capillary Barriers. *Water Resour. Res.* **1993**, *29*, 1045–1056. [CrossRef]
21. Stormont, J.C.; Morris, C.E. Method to Estimate Water Storage Capacity of Capillary Barriers. *J. Geotech. Geoenviron. Eng.* **1998**, *124*, 297–302. [CrossRef]
22. Rahardjo, H.; Tami, D.; Leong, E.C. Effectiveness of sloping capillary barriers under high precipitation rates. In Proceedings of the 2nd International Conference on Problematic Soils, Selangor, Malaysia, 3–5 December 2006; pp. 39–54.

23. Scarfone, R.; Wheeler, S.J.; Lloret-Cabot, M. A hysteretic hydraulic constitutive model for unsaturated soils and application to capillary barrier systems. *Géoméch. Energy Environ.* **2020**. [CrossRef]

24. Scarfone, R.; Wheeler, S.J.; Lloret-Cabot, M. Conceptual Hydraulic Conductivity Model for Unsaturated Soils at Low Degree of Saturation and Its Application to the Study of Capillary Barrier Systems. *J. Geotech. Geoenviron. Eng.* **2020**, *146*, 04020106. [CrossRef]

25. Santacroce, R.; Cioni, R.; Marianelli, P.; Sbrana, A.; Sulpizio, R.; Zanchetta, G.; Donahue, D.J.; Joron, J.L. Age and whole rock–glass compositions of proximal pyroclastics from the major explosive eruptions of Somma-Vesuvius: A review as a tool for distal tephrostratigraphy. *J. Volcanol. Geotherm. Res.* **2008**, *177*, 1–18. [CrossRef]

26. Sepe, C.; Calcaterra, D.; Di Martire, D.; Ramondini, M.; Russo, G.; Vitale, E.; Pappalardo, L. Role of capillary barriers in landslide susceptibility assessment in pyroclastic soils: Microstructure investigation and numerical analysis. *Environ. Earth Sci.* **2021**, unpublished.

27. De Vita, P.; Di Clemente, E.; Rolandi, M.; Celico, P. Engineering geological models of the initial landslides occurred on april 30 2006, at the Mount of Vezzi (Ischia island, Italy). *Ital. J. Eng. Geol. Environ.* **2007**, *2*, 119–141.

28. Cruden, D.M.; Varnes, D.J. Landslide types and processes. In *Landslides: Investigation and Mitigation*; Turner, A.K., Schuster, R.L., Eds.; Transportation Research Board Special Report 247, National Research Council; National Academy Press: Washington, DC, USA, 1996; pp. 36–75.

29. Bertagnini, A.; Landi, P.; Rosi, M.; Vigliargio, A. The Pomici di Base plinian eruption of Somma-Vesuvius. *J. Volcanol. Geotherm. Res.* **1998**, *83*, 219–239. [CrossRef]

30. Pappalardo, L.; Buono, G.; Fanara, S.; Petrosino, P. Combining textural and geochemical investigations to explore the dynamics of magma ascent during Plinian eruptions: A Somma–Vesuvius volcano (Italy) case study. *Contrib. Miner. Pet.* **2018**, *173*, 61. [CrossRef]

31. Buono, G.; Pappalardo, L.; Harris, C.; Edwards, B.R.; Petrosino, P. Magmatic stoping during the caldera-forming Pomici di Base eruption (Somma-Vesuvius, Italy) as a fuel of eruption explosivity. *Lithos* **2020**, *370*, 105628. [CrossRef]

32. Arya, L.M.; Paris, J.F. A physicoempirical model to predict the soil moisture characteristic from particle size. *Soil Sci. Soc. Am. J.* **1981**, *45*, 1023–1030. [CrossRef]

33. Cecconi, M.; Vecchietti, A.; Pane, V.; Russo, G.; Cencetti, C. Geotechnical Aspects in the Assessment of Stability Conditions of the Volumni Hypogeum in Perugia. In *Geotechnical Research for Land Protection and Development, CNRIG 2019*; Calvetti, F., Cotecchia, F., Galli, A., Jommi, C., Eds.; Lecture Notes in Civil Engineering 2020; Springer: Berlin/Heisenberg, Germany, 2020; Volume 40.

34. Cecconi, M.; Cambi, C.; Carrisi, S.; Deneele, D.; Vitale, E.; Russo, G. Sustainable Improvement of Zeolitic Pyroclastic Soils for the Preservation of Historical Sites. *Appl. Sci.* **2020**, *10*, 899. [CrossRef]

35. Van Genuchten, M.T. A closed-form equation for predicting the hydraulic conductivity of unsaturated soils. *Soil Sci. Soc. Am. J.* **1980**, *44*, 892–898. [CrossRef]

36. Olivella, S.; Carrera, J.; Gens, A.; Alonso, E.E. Non isothermal multiphase flow of brine and gas through saline media. *Transp. Porous Media* **1994**, *15*, 271–293. [CrossRef]

37. Olivella, S.; Gens, A.; Carrera, J.; Alonso, E.E. Numerical formulation for a simulator (CODE_BRIGHT) for the coupled analysis of saline media. *Eng. Comput.* **1996**, *13*, 87–112. [CrossRef]

38. Mualem, Y. A new model for predicting the hydraulic conductivity of unsaturated porous media. *Water Resour. Res.* **1976**, *12*, 513–522. [CrossRef]

39. Lu, N.; Likos, W.J. *Unsaturated Soil Mechanics*; Wiley: New York, NY, USA, 2004; ISBN 978-0-471-44731-3.

40. Bilotta, E.; Cascini, L.; Foresta, V.; Sorbinow, G. Geotechnical characterisation of pyroclastic soils involved in huge flowslides. *Geotech. Geol. Eng.* **2005**, *23*, 365–402. [CrossRef]

41. Vanapalli, S.K.; Fredlund, D.G.; Pufahl, D.E.; Clifton, A.W. Model for the prediction of shear strength with respect to soil suction. *Can. Geotech. J.* **1996**, *33*, 379–392. [CrossRef]

MDPI
St. Alban-Anlage 66
4052 Basel
Switzerland
Tel. +41 61 683 77 34
Fax +41 61 302 89 18
www.mdpi.com

Geosciences Editorial Office
E-mail: geosciences@mdpi.com
www.mdpi.com/journal/geosciences